NAVIGATING
WITH WHITE-FACED
CAPUCHIN MONKEYS

sidestonepress

NAVIGATING
WITH WHITE-FACED
CAPUCHIN MONKEYS

Primate Behavioral Ecology and Spatial
Cognition in a Mesoamerican Rainforest

Bernardo Urbani

Sidestone Press Dissertations
Published by Sidestone Press, Leiden
www.sidestone.com

Lay-out & cover design: Sidestone Press
Photograph cover: White-Faced Capuchin Monkey
(Mariana Ianovska, stock.adobe.com)
Illustration back: White-Faced Capuchin Monkey from Charles
d´Orbigny´s *Dictionnaire universel d'histoire naturelle. Atlas. Zoologie.*
Tome Premier (1849, Paris. Éditeurs Renard, Martinet et Cie).

ISBN 978-94-6428-057-9 (softcover)
ISBN 978-94-6428-058-6 (hardcover)
ISBN 978-94-6428-059-3 (PDF e-book)

Contents

Acknowledgements   **9**

1. Introduction   **13**

Objectives, Questions, and Hypotheses of this Research   15

Capuchin Monkeys   17

Overview of the Monograph   20

2. Navigating the Literature: A Theoretical Assessment and a Review on Spatial Mapping in Primates   **23**

Types of Mental Maps   24

Exploring Spatial Memory in Wild Primates   28

Examining Spatial Representations in Capuchins   31

3. Study Site, Study Group, Materials, and Methods   **37**

Study Site   37

Study Group   44

Data Collection   46

Research Design: Natural Field Study (Behavioral-Ecological Phase)   46

Research Design: Experimental Field Study (Field Experimental Phase)   49

Data analysis   52

Pilot study   52

Ethical statement   53

Data availability   53

4. The Behavioral Ecology of a Group of Wild White-Faced Capuchin Monkeys (*Cebus imitator*)   **55**

Materials and Methods   59

Results   60

Activity budget   60

Diet, feeding, and foraging behavior   62

Resting sites   70

Ranging patterns   72

Discussion 73

    Activity budget 74

    Diet, feeding, and foraging behavior 74

    Ranging patterns 77

5. Spatial Mapping in Wild White-Faced Capuchin Monkeys **87**
(*Cebus imitator*): A Natural Field Study

    Results 87

        Pattern of feeding/resting tree visits and distribution 87

        Canopy density and forest profile: Field-of-view in the forest canopy 92

        A test of spatial memory: The case of the use of two feeding tree species 95

        Evaluating travel itineraries 96

        Use of nodes and route segments 100

        Reaching the major feeding/resting trees from different directions 103

    Discussion 105

6. Spatial Mapping in Wild White-Faced Capuchin Monkeys **111**
(*Cebus imitator*): An Experimental Field Study

    Material and Methods 111

    Results 114

        Experiment 1 114

        Do white-faced capuchins visit experimental feeding platforms in 114
the same order as they were first encountered?

        Do white-faced capuchins travel between sequential feeding 117
platforms using a distance-minimizing principle, and over the
course of the field experiment do they reduce their circuity index
in traveling between experimental platforms?

        Do white-faced capuchins use traditional routes of travel to reach 119
experimental feeding platforms or do they select novel travel
routes?

        Do white-faced capuchins select nearer platforms over more distant 120
platforms?

        Do white-faced capuchins exhibit a win-shift foraging rule when 121
selecting experimental feeding platforms?

        Experiment 2 121

        Do white-faced capuchins visit experimental feeding platforms in the 121
same order as they were first encountered?

        Do white-faced capuchins travel between sequential feeding platforms 123
using a distance-minimizing principle, and over the course of the field
experiment do they reduce their circuity index in traveling between
experimental platforms?

        Do white-faced capuchins use traditional routes of travel to reach 125
experimental feeding platforms or do they select novel travel routes?

Do white-faced capuchins select nearer platforms over more   126
distant platforms? Do white-faced capuchins travel to more distant feeding
platforms that contain higher food rewards preferentially
over nearer feeding platforms that contain lower food rewards?

Do white-faced capuchins exhibit a win-shift foraging rule when   128
selecting experimental feeding platforms?

Discussion   128

7. Conclusions   **133**

References   **141**

Appendix: The Study Area at La Suerte Biological Field Station (EBLS)   **163**

A las personas que creen en los soñadores...
To the people who believe in dreamers...

To my father, Franco Urbani.
To my mentor, Paul A. Garber

Acknowledgements

I would like to thank my family, especially Ana María and Lucía as well as my mother, sister, and nephews. My father deserves special acknowledgment, and actually, this monograph is dedicated to him. I especially thank Paul A. Garber, my former advisor, for his teachings and friendship. I also want to thank Paul and Chrissie for their help in the field during the initial part of the experimental phase. I appreciate Steve Leigh, Susan Ford, and John Polk for their comments, support, and sharing their knowledge. To Barry Lewis for introducing me to the world of computerized mapping.

In Costa Rica, I want to express my gratitude to the Molina family for their support, particularly Renee and Alvaro Molina. At the Estación Biológica La Suerte special thanks to Israel Mesen-Rubí and Raquel Mesen-Rubí de Madrigal, and overall to Jonathan Mesen-Rubí. Also in the field, I really appreciated the friendship and collaboration of Jhonny Cambronero-Blanco, Austín Umaña-Rueda, Hellen Masis-Solano, Elías Mesen-Alemán, Ramón Mesen-Alemán, Cecilia Rubí-Elizondo de Mesen, Isabel Moya-Lobo de Díaz, Marcelino Díaz-Díaz, Luis "Papelo" Díaz-Ledezma, Marcos "Chito" Díaz-Ledezma, Fernando "Pacheco" Díaz-Ledezma, Félix Ramón Álvarez-Malespín, Mauren Mesen-Rubí, Yeltin Vallejos-Leiva, Yeiner Solis-Rubí, Ligia de Cubillos, Yanci Agüero-Rubí, Alfonzo Madrigal-Jarquín, Meraris, Kendall Mesen-Rubí, Andreidyn Elizondo-Leiva, Roberto "Pilo" Cubillos-Mendoza, Marcedonio López-López, Frankis Umaña-Rueda, Sr. Quintero, Jesús Madrigal-Jarquín, and Berni Carranza. At the station thanks also go to Deborah Curtis, Hal Smith, Michelle Bezanson, and Sarah Y. Smith. Special thanks to Ricardo Vázquez at the Museo Nacional de Costa Rica for his cooperation and sponsorship, Reinaldo Aguilar (INBIO) for his invaluable collaboration in botanical identification, as well as Gabriela Bonilla in San José.

At UIUC, I really want to thank Martín Kowalewski, my great friend. My extended gratitude to all my new friends, Rodolfo Martínez-Mota and Nicoletta Righini, Carolina Sternberg, Greg and Rachel Blomquist, Jodi Blumenfeld and Andrew O'Baoill, Sara Ortiz-Escalante and Jordi Honey-Roses, Kanako Iuchi, Batamaka Some and family, Jin-Heon Jung and family, Will and Tina Hope and family, Melissa and Rob Schofield-Raguet, Robin Bernstein, Petra and Germán Bollero-Jelinek and family, Kellie and Justin Glessner and family, Caie Yan, Steve Maas and Sarah Rowe, Eva Pajuelo, Margaret Brown, Angelina Cotler, Pilar Eguez, Nicole Tami, Julián Norato and family, Eduardo and Adriana Herrera-Cuervo, Katie O'Brien, Isabel Scarborough, Soo Jin Park, Jeniffer Hardin, Jennifer Shoaff,

Joy Sather-Wagstaff, Sujey Vega, Donna White, Jill Wightman, Alison Goebel, Michelle Wibbelsman, Karin Berkhoudt, Matthew Anderson, Hal Fischer and Alison Rode, Jun Wan, Betsy Beymer, Erguin Bulut, Sarah Y. Smith, Krista Milich, Scott Williams and Milena Shattuck, Lance Larkin, Talia Melber and Mark Grabowski (in Carbondale: Chihiro Shibata, Juan Luis Rodríguez and Matthew Nowak). To all anthropology professors, especially Stanley Ambrose, Rebecca Stumpf, Barry Lewis, Alejandro Lugo, Andrew Orta, Kathryn Clancy, Ripan Malhi, Matti Bunzl, Christina Grassi, and Leslea Hlusko. Also on campus, thanks to my Venezuelan friends Lourdes González, Luis Miguel Vásquez, and Dora and Luis Salazar. Thanks to the library staff of the University of Illinois at Urbana-Champaign (UIUC) for their aid in reference searching. Outside UIUC, many primate people were there to support me during the development of my student times and particularly during my very first primatological steps. They are, Marilyn Norconk, Julio César Bicca-Marques, Anthony Rylands, Elisabetta Visalberghi, Robert Sussman†, Loretta Cormier, Gabriel Zunino, Alejandro Estrada, Juan Carlos Serio-Silva, Manuel Lizarralde, among others, and later, Dionisios Youlatos, Janet Browne, Carlos Serrano-Sánchez, and Eckhard W. Heymann. To all of them, thanks a lot. Today, to the people at the Behavioral Ecology and Sociobiology Unit of the Leibniz Institute for Primate Research/ German Primate Center, especially, E. W. Heymann, as well as Sofya Dolotovskaya, Peter Kappeler, Claudia Fichtel, Elif Karakoc, Gabriel Robinson-González, Simon Brettschneider, and Christina Glaschke. Also in Göttingen, thanks to my friend Julián Padró.

In Caracas, thanks to the people at EGAL (Geociencias & GIS), particularly Mustapha Boujana, Margiory Marcucci, and Antonio Marcucci for their extreme cooperation in ArcGIS™ work. Also to Enzo Caraballo and Victor Cano for their help with the ArcGIS™ part of this project, and Katiuska Velázquez for her cooperation. Thanks to Pablo Quijada and Alí Gómez for their help with AutoCAD™ and CorelDRAW™, respectively. In Iowa City, thanks to María Alejandra Pérez for her excellent proof-editing cooperation, and friendship. In Argentina, to Vanina Férnandez for sharing her figure. To my extended family the Urbani, Nouel, Oliviero, Piemonte, and Schellongowski. To my old friends, Diego Urdaneta and Johann Starchevich as well as Vanessa Briceño and Paula Sarco-Lira. To the friend at the UCV: Krisna Ruette, Marcia López, Yoli Velandria, Magda Duarte, Leticia Marius, Nuria Escobar, Alejandra Martínez, Juan Carlos Rey, Johan Rodríguez, Diego Delgado, Laura Perozo, Yadira Rodríguez, Héctor Cardona, César Marín, Ronna Villalba, Nicolas González, Francia Medina, Cristina Soriano and brother Juan Luis Rodríguez. Special thanks to the people of the Sociedad Venezolana de Espeleología, with them I learn how to be in the field; they are Joris Lagarde, Francisco Herrera, Ángel Viloria, Carlos Bosque, Joaquín Astort, Juan Antonio Tronchoni, Elizabeth Ohep, Khalil Gheim, Pedro Ascanio, Pedro Aso, Miguel Ángel Perera†, Enrique Bolón, Sheila Marques, Juan Nolla, Jesús Otero as well as Franz Scaramelli and Rafael Carreño. To my Venezuelan primate colleagues Natalia Ceballos-Mago, Carlos Portillo Quintero, Mailén Riveros, Antmar Herríquez, Cristina Gomes, and Carmen Ferreira. At the Universidad Central de Venezuela, thanks to Kay Tarble, Luis Molina, Rosario Massimo, Jesús Oyalbi†, Valentín Fina†, Emmanuele Amodio, Miguel Ángel Perera†, Gustavo Martín†, and Adelaida Struck. At the Instituto Venezolano de Investigaciones Científicas to Erika Wagner, Lilliam Arvelo, Horacio Biord, Hortensia Caballero, Werner Wilbert, Rafael Gassón, Franz Scaramelli, Alberta Zucchi†, Berta Pérez,

Stanfort Zent, Eglee López-Zent, Eliécer Arias, Abel Perozo, Ángel Viloria, Ascanio Rincón, Astolfo Mata, and Francisco Herrera, and Carlos Bosque at the Universidad Simón Bolívar.

I would like to thank the Fulbright-OAS Ecology Initiative Program (LASPAU), the Graduate College-UIUC, and the Department of Anthropology for supporting and funding my graduate studies. To the members of the Fulbright Commission in Venezuela (María Eugenia Méndez and Mariangélica Palma) and in the US (Renee Hahn and Christina Korinek). Thanks to Liz Spears, Karla Harmon, Shari French, Julia Spitz, and Donna Fogerson, who made my academic life at UIUC so much easier.

I really appreciate the institutions that provided funding for this project. In a pre-dissertation stage: the Beckman Foundation/Beckman Institute for the Advanced Science and Technology-UIUC (twice), the Tinker Foundation/Center for Latin American and Caribbean Studies-UIUC, and the Department of Anthropology-UIUC. During the dissertation field research phase: the National Science Foundation (NSF BCS #0612771), the Graduate College-UIUC, the American Society of Primatologists, the Idea Wild Foundation, the Department of Anthropology-UIUC, and the American Philosophical Society. At the time of the submission of this monograph, I appreciate the support of the Alexander von Humboldt Foundation. To all of them, thank you for trusting that I could accomplish this research.

1

Introduction

A central issue in biological anthropology involves the understanding of primate cognition and how prosimians, monkeys, apes, and humans store, encode, represent and integrate spatial and ecological information (Janson 1998, Milton 2000, Garber 1989, 2000, Bicca-Marques and Garber, 2004, 2005, Dolins and Mitchel 2010). Living non-human primates exploit resources within home ranges that vary in size from 0.5-one hectare in pygmy marmosets (Soini 1993) to several hundreds of hectares in chimpanzees (Herbinger *et al.* 2001), and therefore individual species face different challenges associated with tracking the spatiotemporal availability and location of food resources (Tomasello and Call 1997, Call 2000, Garber 2000, Bicca-Marques and Garber 2005, Byrne and Janson 2007, Garber and Dolins 2014). These challenges include the ability to recall the locations of productive feeding sites, the selection of efficient travel pathways between distant targets or goals, and the use of predictable features in the environment such as forest borders, river crossings, or natural topography to navigate within the forager's home range.

In the case of early humans, for example, the coordination of group-based activities to locate and exploit feeding sites and lithic sources that span distances of several kilometers across a range of forested and opened landscapes may have played an important role in the evolution of increased brain size, tool use, foraging strategies, and cognitive ability (Mann 1981, Potts 1998, 2004, Leonard and Robertson 2000, McCabe 2000, Ambrose 2001, Mitchen 2003, Boehm 2004, Wynn 2010). In this regard, a critical evaluation of spatial navigation in non-human primates offers a productive framework to examine decision-making and the set of ecological and behavioral factors that may have shaped the evolution of cognitive abilities in human ancestors (Milton 1981, 1988; Barton 2000; Bicca-Marques and Garber 2005).

In this book, I examined spatial mapping and decision-making in wild white-faced capuchin monkeys (*Cebus imitator*). In particular, I addressed a series of questions concerning primate foraging strategies and the ability of New World monkeys to integrate spatial, temporal, and quantity information to locate feeding sites. In order to examine how capuchins mentally represent spatial information and the kinds of information used in foraging decisions, I collected data integrating two research approaches: natural field studies and experimental field studies (see details in Chapter 3). Natural field studies offer the opportunity to document the specific

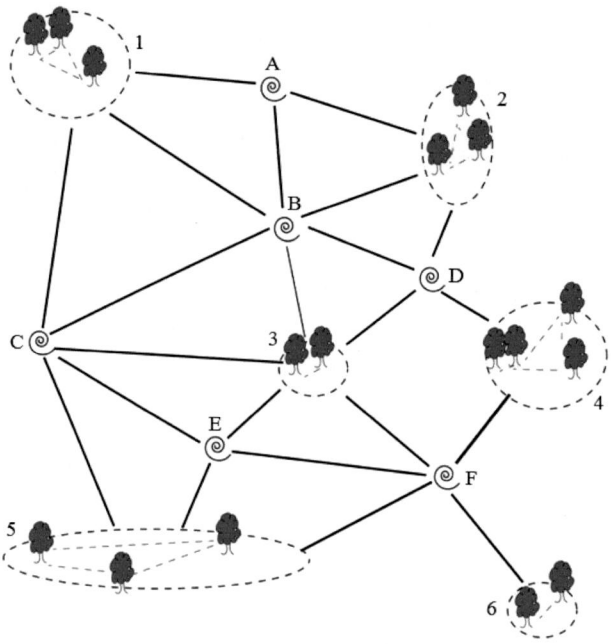

Figure 1.1. Graphic representation of the Poucet Model. Lines are "routes." Feeding/resting trees are within the dot lines. Spirals are "nodes" (After Poucet [1993] and Garber [2000]. Adaptation courtesy of V. Fernández, from Fernández [2008]). For example, using a route-based spatial representation a forager traveling from point 1 to point 6 might go through a set of nodes (e.g. spirals C, E, and F) to reach its final goal. In contrast, using a coordinate-based spatial representation, a forager traveling from 1 to point 6 might compute a direct and straight route between both points.

set of foraging problems an individual encounters, and the precise travel routes, patterns of habitat utilization, and decisions made to solve these problems. Natural field studies are limited; however, by the inability of the researcher to control the specific information available to the forager, and therefore to directly test hypotheses concerning primate decision-making (Garber and Lavalle 1999, Garber and Bicca-Marques 2007, Janson and Brosman 2013). Therefore, I also included field experiments as part of this research. Experimental field studies complement both naturalistic and laboratory research by presenting wild primates with ecological and social problems analogous to those they naturally encounter but under more controlled conditions (Janson and Di Bitteti 1997, Janson 1998, Garber and Lavalle 1999, Garber 2000, Di Bitteti and Janson 2001, Bicca-Marques and Garber 2005, Garber and Bicca-Marques 2007, Janson and Brosman 2013). The combination of approaches used in this research (i.e. observing the natural foraging and spatial navigation of wild capuchins and then conducting field experiments on foraging decisions and use of travel routes under more controlled conditions by systematically varying food availability and distances between feeding platforms) offers a set of critical methodological tools in addressing questions of spatial navigation, decision-making, and cognition in wild primates.

In this research, I tested a series of hypotheses concerning the degree to which capuchin monkeys rely on a route-based spatial representation or a coordinate-based spatial representation to integrate ecological information. Using a route-based map, a primate is expected to integrate the spatial location of travel routes with landmarks and topographical features of the environment such as emergent trees, elevated terrain, or forest and river borders (Byrne 2000, Garber 2000, Valero and Byrne 2007, Di Fiori and Suárez 2007, Abreu et al. 2021). For example, a forager using a route-based spatial representation is expected to navigate utilizing a relatively small set of predictable landmark cues distributed across their home range. These landmarks act as nodes or switch points from which the forager re-orients travel direction and reuses a set of familiar arboreal pathways. Using a route-based spatial representation, the forager is not expected to travel the shortest distance or select the most direct route to reach its goal. Rather it is expected to use and reuse a set of familiar routes of travel. In contrast, using a coordinate-based spatial representation a primate is expected to encode spatial information in the form of a coordinate-based map where elements of the environment are represented as precise distances and directions with regard to fixed points in the environment (Byrne 2000, Garber 2000). A forager using a coordinated-based system is expected to compute novel, straight-line, and direct travel routes to reach its goal. Poucet (1993) has argued that foragers may rely on both types of spatial representations to travel to feeding and resting sites (Fig. 1.1). In his model, a forager is expected to use a route-based spatial representation in large-scale space to reach a distant feeding or resting site, and use a coordinate-based spatial representation in small-scale space to locate their goal. Differences in navigating in large- and small-scale spaces are associated with the ability of the individual to obtain and encode views of the same set of landmarks from multiple directions.

Objectives, Questions, and Hypotheses of this Research

The primary objective of this study is to test a series of hypotheses concerning spatial cognition and decision-making in wild Costa Rican white-faced capuchins (*Cebus imitator*) in large- and small-scale spaces. Some of the major questions addressed in this research include:

1. Are capuchin ranging and travel patterns consistent with a route-based or a coordinate-based spatial representation?
2. Do capuchins orient to particular landmarks in the environment during long-distance travel?
3. When presented with a set of experimental feeding platforms that vary in spatial distribution, the distance between adjacent platforms, and amount of the food reward, do capuchins, (a) visit platforms using the most direct or efficient route (least distance)?, (b) travel to nearer platforms that contain equal or a reduced reward or (c) select more distant platforms that contain a higher food reward (prioritize distance over food reward)?.

Hypotheses (H)	Data collected	
H_1: If capuchins use a coordinate-based spatial representation, then they are expected to	$H_{1.1}$) travel to trees in a relatively straight-line fashion, using shortcuts or direct novel routes of travel.	a) calculating the circuity index between trees and/or foraging patches visited at different distances.
H_2: If capuchins use a route-based spatial representation, then they are expected to	$H_{2.1}$) travel to previously visited trees by re-using a set of travel route segments.	b) recording the number of feeding and resting trees visited per day, and how frequently these sites were revisited over the course of a single day, consecutive days per month, and months.
	$H_{2.2}$) utilize particular forest places of the forest such as nodes as switch points to re-orient travel.	b) recording the frequency of re-use of travel route segments.
		d) recording whether particular physical features of the groups' home range such as nodes or switch points were used by capuchins to re-orient travel.
		e) in two fruiting tree species (*Dipteryx panamensis* and *Sterculia recordiana*), recording whether nearer trees of that species were selected over more distant trees of that species as the next feeding site.
		f) calculating an estimated "field-of-view" or visibility index from the most commonly used feeding and resting sites.
		g) recording whether major feeding/resting trees were reached from different directions.
H_3: During Experiments 1 and 2, if capuchins primarily make foraging decisions based on distance rather than the amount of food reward, then	- When the quantity of the food reward is identical on platforms located at similar distances then, $H_{3.1}$) the capuchins are expected to visit nearer feeding sites once a previous feeding platform was visited.	a) recording the order in which feeding platforms were discovered.
		b) comparing the order of platform discoveries with the spatial information recorded during the natural field study (between March and August)
	- At different distances and different amounts of food reward, it is expected that capuchins, $H_{3.2}$) visit the nearer feeding site even when this involves expectations of lower food reward.	c) calculating and recording the sequence of platforms' visits during square and rectangular spatial configurations.
		d) recording whether nearer feeding platforms were selected over more distant platforms as the next feeding site.
H_4: – If capuchins use a route-based spatial representation, then during field experiments: $H_{4.1}$ – If capuchins use a coordinate-based spatial representation, then during the same field experiments: $H_{4.2}$	$H_{4.1}$) -it is expected that they will continue to re-orient travel at commonly used nodes to reach feeding platforms even when these routes are characterized by increased travel distance. – it is expected to have higher circuity indices while traveling to feeding platforms.	a) comparing whether the nodes used during the natural field study (between March and August) were used during the experimental field study.
		b) calculating the circuity index of the routes of travel used to reach the platforms at both large-scale space and small-scale space.
	$H_{4.2}$) -they are expected to compute novel short-cuts and more direct routes between experimental feeding sites. – it is expected to have lower circuity indices while traveling to feeding platforms.	c) calculating the degree to which routes used during the natural field study (between March and August) were also used during the experimental field study.
		d) examining the learning curve of the monkeys' movement to feeding platforms using the circuity indices calculated in each experiment.

Table 1.1. Hypotheses tested and data collected in this study.

The first phase of the project was an 8-month natural or behavioral-ecological field study of a group of ~12 habituated white-faced capuchin monkeys inhabiting La Suerte Biological Station (EBLS) in northeastern Costa Rica. I documented ranging behavior, diet, and the spatiotemporal patterns by which white-faced capuchins reached sequential feeding and resting sites. In the second phase of this research, I conducted two experimental field studies in which the spatial location, the distance between feeding platforms, and the amount of food reward on feeding platforms were systematically varied. During both phases of this research, I tested a set of hypotheses. These hypotheses and the data collected to test them are outlined in Table 1.1.

Capuchin Monkeys

Capuchins are taxonomically divided into two major groups. The "tufted" or "robust" capuchin monkeys (genus *Sapajus*) include, among others, the black-horned or black capuchin (*S. nigritus*), the black-striped or bearded capuchin (*S. libidinosus*), the brown capuchin (*S. apella*), and the critically endangered golden-bellied or yellow-breasted capuchin (*S xanthosternos*) (Groves 2001). The "untufted" or "gracile" capuchin monkeys (genus *Cebus*) include different species such as the wedged-capped or weeper capuchin (*C. olivaceous*, formerly known as *C. nigrivittatus*), the Ka'apor or Kaapori capuchin (*C. kaapori*), the white-fronted capuchin (*C. albifrons*), and the white-faced capuchins (*C. imitator* and *C. capucinus*) (Groves 2001). The *imitator* form inhabits the north Atlantic coast of Costa Rica, and was observed during this study (Fig. 1.2).

Except for howler monkeys (genus *Alouatta*), capuchins exhibit the most geographically widespread distribution among Neotropical primates, ranging from Honduras to northern Argentina (Wolfheim 1983, Groves 2001). White-faced capuchins are the only capuchin species present in Mesoamerica; with a distribution from Honduras to the Pacific coast of Colombia and Ecuador in South America (Defler 2004). Capuchins inhabit a variety of forest types including primary and secondary forests, highly disturbed forests, seasonal forests, mangroves, swamps, dry forests, rain forests, cloud forests, semi-deciduous forests, gallery forests, flooded forests, and forest tracts in *llanos*, *caatinga*, and *cerrado* environments (Hernández-Camacho and Cooper 1976, Freese and Oppenheimer 1981, Wolfheim 1983, Robinson and Janson 1987, Fragaszy *et al.* 2004, Defler 2004, Jack 2007). These areas vary in altitude from sea level to 2700 m a.s.l. (Hernández-Camacho and Cooper 1976, Linares 1998, Defler 2004). Capuchin home ranges vary in area from 65 to 115 ha in *C. albifrons*, 32.5 ha to 36 ha to 358 ha in *C. imitator* (see dataset and references in Chapter 4: Table 4.23). The mean group size is 19.3 individuals/group in *C. imitator*. In Chapter 4 (Tables 4.21, 4.22, 4.23), I present data on activity budget, diet, ranging patterns, and group size in wild untufted capuchins.

Captive white-faced capuchins weigh an average of 3.3 kg and are the heaviest capuchin species (Ford and Davis 1992). Ford and Davis (1992) indicate that males weigh approximately 33% more than females (♂=3,868 g, ♀=2,666 g). Compared to other New World primates, capuchins have relatively long life spans (Fragaszy *et al.* 2004). A study suggests that wild white-faced capuchins may live up to 30 years (MacKinnnon 2006). Capuchins have an extended juvenile period (MacKinnon 2002). The age at first birth in wild capuchins is ~7 years, and males reach reproductive maturity at 10 years (Robinson and Janson 1987, Fedigan and Rose 1995, Kappeler and Pereira 2003, Fragaszy *et al.* 2004,

Figure 1.2. White-faced capuchin monkeys (*Cebus imitator*) at La Suerte Biological Field Station (EBLS), northeastern Costa Rica (Photographs by B. Urbani).

Jack and Fedigan 2004, Jack 2007). *Cebus* spp. live in multimale-multifemale groups and are unusual among New World primates in the degree to which females are philopatric. Based on extensive field observations, it appears that dispersal in *C. albifrons*, *C. imitator*, and *C. olivaceus* is male-biased. This is not a common pattern among Neotropical primates, but also is reported in *Saimiri sciureus* (Boinski *et al.* 2005). Female philopatry may be related to the fact that capuchin groups have strong female-female social bonds (Pope 2000, but see *Saimiri sciureus*) and female coalitions (Jack 2007).

Fedigan and Jack (2006) present data on the dispersal pattern in male *C. imitator*. Based on genetic data of 16 males and 19 infants, these authors suggest that both paired male migration (e.g. normally subadult males departing the group) and parallel male migration (e.g. sibling males leaving together from a group) may characterize some white-faced capuchin populations. Female philopatry and male dispersal appear to characterize all or most capuchin species (Urbani and Kowalewski 2021).

Several researchers have argued that capuchin monkeys represent an important animal model for addressing questions of brain size, cognition, and problem-solving in primates (Visalberghi 1990, Fragaszy *et al.* 2004, Garber *et al.* 2008, Urbani and Garber 2002). For their body size, capuchins have brains proportionately larger than chimpanzees (Gibson 1986, Gilissen 1993, 2003). The encephalization quotient in capuchins is 2.54–4.79 (Jerison 1973), which is twice as large as expected for a non-primate mammal of its body size. Capuchins are distinctive among Neotropical primates because of their complex expression of cerebral fissures including a significantly greater amount in the cortex area compared to other platyrrhines (Hershkovitz 1977). In addition, relative to cortex volume, the size of the cerebellum and the thalamus in capuchins is similar to that found among apes (Stephan *et al.* 1988). The cerebellum and the thalamus are suggested to play a role in the integration of sensory information including motor coordination, and the integration of visual and olfactory information (Altman 1997, Steriade *et al.* 1997, Barlow 2002).

Capuchin monkeys have semi-opposable thumbs and fingers that move independently (Spinozzi *et al.* 2004, 2007). They are described as extractive foragers and consume ripe fruits, seeds, nuts, invertebrates, pith, nectar, and occasionally vertebrates (Freese and Openheimer 1977, Janson and Boinski 1992, Fragaszy *et al.* 2004). Wild capuchins have been observed cracking protected and encapsulated food items such as nuts and oysters against hard substrates (Fernandes 1991, Panger *et al.* 2002, Urbani and Garber 2002, Fragaszy *et al.* 2004). Tufted capuchins have thick molar enamel, larger canines, and larger molars than untufted forms, which enables them to masticate hard food items (Kay 1981, Kay *et al.* 1988, Masterson 2003, Wright 2004). Capuchins species, however, differ in feeding ecology, diet, and masticatory anatomy. For example, the jaws and skull of brown capuchins (*S. apella*) are described by having a robust morphology, allowing powerful mastication of tough food items. This distinguishes them from untufted capuchins (*C. olivaceus, C. imitator*, and *C. albifrons*). In this regard, Wright (2003, 2004) has argued that the masticatory anatomy of *S. apella* provides an advantage in exploiting mechanically tough and encased food items such as palm nuts. In the case of *S. libidinosus*, hard nuts are opened not only by their jaws and teeth but also by using stones as tools (e.g. Moura and Lee 2004, Fragaszy et al. 2004). In some populations of *S. libidinosus* in dry habitats in Brazil, tool use is a common behavioral pattern (Moura and Lee 2004, Fragaszy *et al.* 2004, Ottoni and Mannu 2008; for *S. xanthosternos*: Rodrigues-Canale *et al.* 2009). However, based on a series of experimental studies in captivity and in the wild, it remains uncertain the degree to which capuchins possess causal knowledge, or an understanding of how tools function in accomplishing a task (Visalberghi and Limongelli 1994, Panger 1998, 1999, Garber and Brown 2004).

Individual white-faced capuchin groups displayed local dietary and social "traditions" (Fedigan 1990, Perry *et al.* 2003). Traditions are defined as "a behavioral practice that is relatively long-lasting (i.e. is performed repeatedly over a period of time), and that is shared among members of a group in part through social learning" (Perry and Mason 2003: 71). In Costa Rica, Chapman and Fedigan (1990) proposed that the dietary differences found among three neighboring groups of *Cebus imitator* were not explained by the local abundance of food resources. They argued that between group dietary differences, or the selection of different types of plant items that were similarly available to three capuchin groups in all study years, were best explained in terms of social learning and group traditions. Based on 13 years of data, Perry *et al.* (2004) describe cross-site differences in *C. imitator* social behavior including sucking body parts, hand-sniffing, and "games" that they argue were socially learned and transmitted between members of a given study group (Perry *et al.* 2003). The presence of local traditions, independent of genetic or environmental constraints, suggests that socially transmitted behaviors may play a fundamental role in understanding behavioral plasticity and cognitive ability in *Cebus* (Perry and Mason 2003, Perry 2006).

Overview of the Monograph

In this first chapter, I outline the research questions and hypotheses addressed in this monograph and briefly describe the set of spatial search strategies used by primate foragers. I also present basic information on the biology, behavior, and ecology of capuchin monkeys (*Cebus* spp.). In Chapter 2, I review the literature regarding spatial mapping, memory, and navigation in primates. Key terms such as cognition, spatial mapping, foraging, decision-making, and landmarks are defined. Different forms of internal spatial representations in animals are reviewed. A review of spatial mapping in wild primates is presented. In addition, I summarize information on spatial representations in captive and wild capuchins.

In Chapter 3, I describe the study site, the size and composition of the capuchin study group, methods of data collection, and statistical analyses used to evaluate data gathered in the natural field study and the experimental field study phases of this research.

In Chapter 4, information on the activity budget, feeding behavior, and diet of the capuchin study group are presented. The results indicate that based on time spent feeding and foraging, the capuchins' diet consisted of 57% plant parts, and 42.9% animal prey. During the months of March, April, May, June, September, and October, individual trees of *Dipteryx panamensis* and *Ficus americana* were among the three highest-ranked species exploited by the capuchins. The top three species accounted for ≥50% of capuchin plant feeding time each month. Social interactions represent 3% of the group's activity budget. Grooming was the most common social activity (accounting for 61% of social activity). In general, the behavior and feeding ecology of the white-faced capuchins at EBLS remain stable during the year.

In Chapter 5, I examine spatial mapping in a troop of white-faced capuchins monkeys during an 8-month study period. I tested Poucet's (1993) model of spatial representation. The main contention of this model is that in large-scale space, foragers are expected to use route-based information relying on habitual routes and landmark cues, which function as nodes or switch points. Nodes are fixed points in space used to gain bearing and re-orient travel. In small-scale space, however, the forager may encode different views of the same target into a coordinate-based spatial representation and create novel shortcuts to reach these goals. The results of this chapter suggest that capuchins travel to the nearest neighboring tree of a given species. Capuchins keep detailed knowledge of the location of multiple feeding/resting sites within the forest. The monkeys at EBLS used and re-used particular travel route segments and nodes to reach feeding/resting sites. White-faced capuchins travel deviated 42% more than the most direct distance between goals (111±81 m). However, once the capuchins were in the vicinity of feeding/resting sites (30 m), they re-oriented travel and used more direct routes to reach feeding/resting trees. In sum, capuchins appeared to use coordinate-based spatial representation in small-scale space, once they reach the vicinity of targets. In large-scale space, however, the monkeys relied on a route-based map. Nonetheless, the capuchins also displayed a set of patterns such as high circuity indices and reliance on nodes that are not expected in forming a coordinated-based map. The use of non-direct routes of various distances rather than short-cuts between feeding/resting sites enables a forager to gain information on food availability including insect distribution and the fruiting status of feeding trees. If this is the case, then Poucet's model needs to be reconsidered to include resource sampling in primate decision-making. In this regard, the use of space and routes of travel reveal both the sensory/cognitive ability of the forager and the ecological problems foragers face as well as the most efficient ways of solving these challenges.

In Chapter 6, I examine spatial memory in wild capuchins monkeys during an experimental field study. I found that capuchins visited feeding platforms using a win-shift foraging rule. This means that once they exploited a feeding site, it was no longer part of a set of feeding choices for that day. In addition, the capuchins generally were found to select the nearest platform regardless of the amount of the food reward (four bananas *vs.* eight bananas). In the field experiments, the monkeys re-use nodes that were commonly used during the natural field study. However, the capuchins did not form novel shortcuts when traveling between feeding platforms. I also found that when traveling between experimental platforms, the capuchins did not use straight-line or distance-minimizing travel routes. This suggests that they integrated information as a route-based spatial representation in large-scale space. As in the natural field study, the monkeys used more direct routes once they approached the feeding platforms (30 m); thus capuchins appeared to use a coordinate-based spatial representation in small-scale space. Nevertheless, the results of this chapter do not conclusively support the contention that white-faced capuchins formed a coordinate-based map.

In Chapter 7, I summarize the major findings of this field study and their implications for understanding spatial cognition in non-human primates. As the most significant contribution of this monograph, I found that white-faced capuchins were able to encode and integrate spatial information at different spatial scales. In large-scale space, the monkeys relied on a set of common route segments and nodes and formed a route-based system for spatial navigation. Once the capuchins were in the vicinity of feeding/resting trees, they took highly direct routes to reach targets, suggesting that in small-scale space they plausibly form a coordinate-based spatial representation. In general, these results are consistent with Poucet´s (1993) model of spatial representation. The degree to which similar spatial representations are present in other primate taxa remains unclear, and represents a research challenge that requires further study. On the other hand, the evidence is consistent with using route-based spatial representations at both spatial scales. It may be more efficient, -and with low energetic costs,- for foragers to take multiple travel routes with higher deviations rather than direct shortcuts between sites in order to monitor ecological information. In addition, in small-scale space, capuchins may use near-to-site visual cues to reach their goals. If this is correct, then the model proposed by Poucet (1993) may be too mechanistic in explaining how foragers use particular spatial representations at different spatial scales.

2

Navigating the Literature: A Theoretical Assessment and a Review on Spatial Mapping in Primates

Field studies on wild non-human primates suggest that several species of monkeys and apes travel directly to distant feeding and resting sites, and maintain detailed spatial representations of the distribution of resources in their home range (e.g. New World primates: *Alouatta palliata*, Milton 1980, Garber and Jelinek, 2005; *Ateles geoffroyi*, Milton 1981, Chapman *et al.* 1989, Valero and Byrne 2004; *Ateles belzebuth,* Di Fiori and Suárez 2004; *Cebus apella*, Janson 1990a, 1990b, 1996, 1998; Janson and Di Bitetti 1997; Di Bitetti and Janson 2001; *Cebus imitator*, Garber and Paciulli 1997; *Lagothrix lagotricha*, Di Fiori and Suárez 2004; *Pithecia pithecia*, Cunningham 2003; *Leontocebus fuscicollis*, Garber 1989, 2000, Garber and Hannon 1993, Bicca-Marques and Garber 2003; *Saguinus imperator*, Bicca-Marques and Garber 2003, 2005; *Saguinus mystax*, Garber 1989, 2000, Garber and Hannon 1993; Old World primates: *Papio hamaydras*, Byrne 2000; Apes: *Pan troglodytes*, Boesch and Boesch 1984, Bates and Byrne 2004, Normand and Boesch 2009; and Prosimians: *Eulemur fulvus rufus* and *Propithecus edwardsi*, Erhart and Overdorff 1999, 2008, Overdorff and Erhart 2001). However, little is known about the degree to which primates encode spatial information using a route-based or a coordinate-based spatial representation, and whether primates use a single landmark or a set of landmark configurations in forming spatial maps (Garber and Brown 2006, Normand and Boesch 2009, Urbani this study).

The goals of this chapter are to review the literature regarding spatial mapping and navigation in primates. Explanations of the types of internal spatial representations foragers may maintain are presented. I discuss spatial mapping in primates focusing on spatial representations in captive and wild capuchins. Table 2.1 provided a set of definitions of key concepts addressed in this monograph.

Concept	Definition
Cognition	Animal's internal mechanism of information processing. Integration of previously learned information and/or newly discovered information for understanding the relationship of elements in the individual's environment (Shettleworth 1998, Tomasello and Call 1997, Sternberg 1999, Reznikova 2007).
Spatial mapping	Animal's ability to internally represent, encode and integrate the relative or specific locations of points in the environment, and use information of spatial relationships to reach goals (Gallistel 1990, Wehner and Wehner 1990, Benhamou 1996, Poucet 1993, Cheng and Spetch 1998, Etienne *et al.* 1998, Dyer 1998, 2000, Menzel and Menzel 2006).
Foraging	The act of searching for items that might be acquired and consumed as food (Allaby 1998 in Bicca-Marques 2000)
Decision-making	The process by which an individual chooses one set of information over other sets of information in selecting a behavioral option (Durkas 1998, Shettleworth 1998, Schuck-Paim and Kacelnik 2007).
Landmarks	Fixed points in the environment used during navigation in order to reach a goal. Landmarks can be used simply or as an array. A landmark array is defined as a configuration of beacons that function as associative cues for navigation (Cheng 1986, Braithwaite 1998, Save *et al.* 1998, Sherry 1998, Kamil and Cheng 2001).

Table 2.1. Key concepts of this monograph.

Types of Mental Maps

Based on evidence from controlled laboratory investigations and natural field observations, invertebrates and vertebrates are expected to navigate using both egocentric and/ or geocentric systems of spatial reference (Wehner and Wehner 1990, Benhamou *et al.* 1990). Using *egocentric* coding or path integration (dead reckoning), a forager may track changes in the position of its body relative to the environment and use this information to return to a target or goal (Benhamou *et al.* 1990).

Mauer and Séguinot (1995) suggested that by using path integration an animal can return to its starting point (or home base) from an ending point by reconstructing the bearing, direction, distance, and in some cases the speed of travel used in the outgoing route. Computational action occurs along each segment of the path. The use of landmark information is not required in path integration (Etienne *et al.* 1996).

Examples of path integration include studies on ants and humans. Wehner and Srinivasan (1981) suggest that Saharan ants (*Cataglyphis bicolor* and *Cataglyphis albicans*) forage in a flat open landscape generally lacking landmark information. Based on a series of experimental studies it was found that these ants transport food back to their nest by tracking step-by-step distances, angular changes in body orientation, and speed of travel from a starting point to an endpoint, and then computing a return pathway (Wehner and Srinivasan 1981). In the case of humans, based on field observations, Oatley (1974; see also Farrall 1979) suggested that in the open sea and during the day, Polynesian sailors use path integration to navigate long distances, rather than relying on spatial cues. However, although the sailors know the location of stars, they do not take advantage of stellar configurations for navigation, they avoid traveling at night (Oatley 1974). For navigation, Polynesian sailors maintain a detailed record of distances in relation to travel time between islands. In addition, they make corrections to the speed and bearing of the boats during navigation relative to wind and wave directions and strength (Oatley 1974).

In contrast, a forager using *geocentric* information is expected to rely on environmental cues such as landmarks and natural topographic features of the environment to track its movement in space. A geocentric system is expected to be more efficient for travel in large-scale space (Garber 2000, Byrne 2000) than an egocentric system if the forager can encode, integrate, and compute the relative spatial positions of goals or targets to a single landmark or sets of fixed landmarks in its home range (Cheng and Spetch 1998, Collet and Zeil 1998, Gibson and Kamil 2001, Garber 2000, Garber and Brown 2006).

Landmark navigation occurs when a forager uses one or multiple fixed feature(s) of the environment as a reference point to reach a target or to compute a route between sequential goals (Cheng and Spetch 1998). Pigeons (*Columba livia*) and Mongolian gerbils (*Meriones unguiculatus*) use beacons in experimental tasks to locate and orient direction to particular goals in experimental tasks (Cheng and Sherry 1992, see a detailed experimental example below: Collet *et al.* 1986). It has been argued that the use of multiple landmarks rather than a single landmark can reduce errors in navigation (Cheng and Spetch 1998, Dyer 1998). The use of a configuration of landmarks allows foragers to integrate spatial information relationally and reach goals more efficiently (Braithwaite 1998).

Studies of Mongolian gerbils (*Meriones unguiculatus*) indicate that a landmark or an array of landmarks can be used to find the location of food rewards (Collet *et al.* 1986). These animals were given the task of locating sunflower seeds in relation to visual cues in a black experimental circular arena (diameter 3.5 m) illuminated with a single light bulb. Seeds were hidden within a matrix of black granite chips within the arena. The spatial relationship between the visual cues or landmarks (aluminum cylinders: 11–26 cm diameter, 34–70 cm high) and seeds was at 50 cm and 180° to the individual cylinders. Collet *et al.* (1986) found that gerbils that associated the distance and bearing of food rewards with landmarks successfully located food. When some landmarks were removed or the light was extinguished in the experimental room, there were no visual cues for forming an association with the location of the rewards. Under these circumstances, the searching pattern was performed following the previous navigation experience with the complete set of landmarks. Then, the gerbils formed new trajectories to the goals using the available landmarks. In sum, gerbils used landmarks relationally.

Mental maps are defined as the forager's abilities to form an internal representation of its home range by encoding and incorporating spatial information such as landmarks and goals (see "Spatial mapping" in Table 2.1; Gallistel 1990, Wehner and Wehner 1990, Poucet 1993, Benhamou 1996, Cheng and Spetch 1998, Etienne *et al.* 1998, Dyer 1998, 2000, Menzel and Menzel 2006). A *route-based map* is defined as a spatial representation in which a forager is expected to acquire, recall and integrate a set of interconnected pathways or route segments that are linked by a set of landmarks or nodes (Byrne 1982, Bennet 1996). These landmarks function as switch or choice points and are used to redirect travel and orient to a target (Bennet 1996, Garber 2000). A route-based map is best represented as a network of interconnecting route segments and nodes that are encoded with exaggerated distances and directions between the targets (Garber 2000). This is similar to a subway map in which targets (stations), route segments (railway paths), and intersections (railway crossing points) are represented as exaggerated rather than actual or precise spatial relationships.

The forager using a route-based spatial representation is expected to re-use the same set of pathways to reach resources located in the same general area or part of their range (Suárez 2003). In addition, the forager is expected to rely on predictable elements of the environment such as rivers, forest borders, or emergent trees as beacons for spatial navigation (for humans see: Milton 2000).

As indicated by Byrne (2000), a route-based map is formed as a set of "places" that offer information for navigation. Garber (2000) suggested that using such a spatial framework, a forager might rely on a combination of different route segments and landmarks to navigate to different areas of their home range. A forager using a route-based map is expected to travel in a relatively straight line to reach a switch point, and then turn or reorient travel to reach previously visited feeding and resting sites. Based on observations of humans and baboons, Byrne (1979, 1982) suggested that many primates represent spatial information using a route-based map. Such a map is represented by "traversable routes as [a] network of *strings*, and locations along them as *nodes*. Each string is in effect a program for locomotion: each node, as well as identifying a physical location, may also contain instructions for a change of direction" (Byrne 1982: 246). Using a route-based map a forager is expected to recall and integrate the spatial position of habitual travel routes with particular landmarks and topographical features of its home range and use and reuse this information for navigation. Thus, route-based spatial representations are configured as a set of interconnected route segments with single or multiple landmarks serving as nodes to re-orient travel (Garber 2000).

Coordinate-based maps are spatial representations in which a forager is expected to encode information in the form of true angles and distances in order to compute novel routes or shortcuts to reaching out-of-view goals (Byrne 2000, Garber 2000). A coordinate-based map implies a global metric representation in which the spatial positions of salient features in the environment are stored as "a view from above" (Gallistel 1989, Bennet 1996). Poucet (1993) suggested that in small-scale space (or in the vicinity of the goal) a forager using a coordinate map is expected to encode multiple views of a goal to compute novel shortcut routes. A forager using such a mental representation is expected to visit sites from different previous directions and distances using relatively straight line progression or shortcuts. Backtracking or sharp angular changes in direction are expected to be rare.

Strip maps are spatial representations in which a forager is expected to store information as sequential "snapshots" of the "visual characteristics" of the environment (Collet 1986, Dyer 1996). In order to return to a starting point, the forager matches these "stored images" (snapshots), image by image to retrace the route. The forager "must be able to head in the correct direction along the route, but it needs not to obtain any additional information about its position and direction of movement in the landscape" (Dyer 1996: 87). Baerends (1941) suggested that when bees (*Ammophila campestris*) navigate to and from the hive they fail to integrate information present on either a coordinate-based map or a route-based map. Rather, bees (*Apis dorsata*) rely on an image-matching strategy or strip map to reach their goals (Dyer 1996).

Animals that encode information in route-based maps, coordinate-based maps, and strip maps, all rely on fixed features of the environment to orient in space. Landmarks function as information points. However, the degree and manner in which landmarks are used in each type of spatial representation are different. Using a route-based map the forager needs to encode a set of landmarks that are used as prominent beacons (e.g. nodes and/or topographic features) and located along commonly used paths (Byrne 2000, Garber 2000, Di Fiori and Suárez 2007). The same landmark may be encoded as different views and possibly different points. In contrast, a forager using a coordinate-based map encodes different views of the same landmark as a single point or reference to compute a novel route. Finally, a forager that uses a strip map encodes different views of the same landmark as unrelated snapshots and matches a set of images to reach a goal.

Type of mental maps	Synonyms
Route-based map	-Topological map -Topological mental map -Topology-based mental map -Network map
Coordinate-based map	-Euclidean map -Vector map -Metric map -Geometric map -Coordinate map -Global coordinate-based map
Strip map	-"Snapshots" map

Table 2.2. Types of mental maps and their synonyms.

In sum, a forager internally representing spatial information as a strip map will need to match exact images in its environment in order to navigate. A forager using a coordinate-based spatial representation will travel by computing a relatively straight or direct route to reach goals. A forager using a route-based map will travel along route segments and re-orient travel at a set of commonly used nodes. Table 2.2 summarizes the synonyms of the types of mental maps discussed above as found in the literature.

Finally, Poucet (1993) developed a testable model for examining how foragers navigate and use spatial information in small- and large-scale spaces. ***Poucet's model*** hypothesized that when traveling in small-scale space (e.g. vicinity of a feeding site) or "local dependent area," animals might store spatial information within a coordinate-based system, whereas when traveling in large-scale space (e.g. between distant feeding patches) they are more likely to use a route-based spatial representation. The distinction between small- *vs.* large-scale space is based on the ability of the forager to obtain different views of a target or goal from several points in the landscape.

Poucet (1993) argues that a critical factor in the ability to construct a coordinate spatial representation is the opportunity to obtain views of the same set of landmarks and goals from multiple directions and then to cognitively convert these representations into a metric representation. In tropical forests, such views may be restricted by the denseness of the canopy whereas, in more open habitats, unobstructed views may be available for distances of hundreds of meters (Garber 2000, Dominy *et al.* 2001). A forager using a coordinate-based spatial representation in small-scale space is expected to encode different views of particular targets and landmarks and construct a "view from above," in order to compute novel short-cut routes of travel to reach out-of-view feeding sites. Poucet (1993) argues, however, that when navigating in large-scale space, animals are unable to obtain direct views of goals and landmarks from multiple locations and therefore are expected to employ a route-based spatial representation and to use and re-use a limited set of habitual routes of travel and nodes to reach major feeding/resting sites. These nodes are reliable features of the environment that function as navigation switch points to re-direct travel. Thus, the forager connects route segments by encoding and recalling the spatial position of a common set of landmarks relative to important targets. Foragers can use these cues to reorient travel. A route-based representation restricts the forager to a set of "known" or habitually used travel routes and nodes in going from one target or goal to another. A major question in the study of primate cognitive ecology is the degree to which different primate species form route-based maps and/or coordinate-based spatial maps (Poucet 1993, Garber 2000).

Exploring Spatial Memory in Wild Primates

Traditional approaches to the study of spatial memory in primates have focused on captive experiments in relatively small enclosures. Wild studies examined how primates explore their home ranges, and the travel pattern that they used. Research testing hypotheses of spatial memory in the wild has expanded during the last decades. In Table 2.3, I summarize some major findings of these problem-oriented studies.

Many primate species are reported to reuse path segments or arboreal pathways when traveling in their home range (references in Table 2.3). This pattern of navigation may be more consistent with a route-based representation if these routes are associated with interconnected switching points for re-orientation of travel. However, there is empirical evidence to support the contention that some wild primates may encode spatial information in a coordinate-based system. For example, Boesch and Boesch (1984) argue that chimpanzees (*Pan troglodytes*) encode, track and compute the location of hammer and anvil stones in relation to the location of palm nut (*Panda* and *Coula*) feeding sites and that these apes use travel-limiting routes to locate previously used tools and nut sites.

In a 5-year study of a community of ~70 chimpanzees at Taï National Park (Ivory Coast) in Western Africa, Boesch and Boesch (1984) report that chimpanzees transported hammers to *Panda* trees. In 40% of cases, hammer stones and feeding sites were out of view. In 95% of the cases, hammer stones appeared to be selected by their distance to the feeding sites. Traveling between the location of hammer stones and feeding was in a relatively-straight line. Based on these data, Boesch and Boesch (1984) conclude that chimpanzees recalled the spatial distribution of hammers and feeding sites, and compute shortcuts for reaching their goals. This is consistent with a coordinate-based spatial representation (Boesch and Boesch 1984).

Normand and Boesch (2009) found that when moving between feeding sites, chimpanzees traveled using straight-line path segments. Using a linearity index (direct distance/distance traveled where 1 = straight line and 0 = backtracking), the authors found that chimpanzees traveled only 4% greater than the most direct distance (linearity index average = 0.96/1). In testing the hypothesis that chimpanzees may represent their space using a coordinate-base map, the authors did not find significant differences in the linearity index between the route segments traveled in the periphery of the home range (used for 25% of their active time) and those used in the core area (used 75% of the time). Thus, the authors concluded that chimpanzees use this type of spatial representation throughout their complete home range. Moreover, the chimpanzees were observed to reach the same feeding trees by traveling from different directions. This suggested that the chimpanzees may have formed a coordinate-based spatial representation in relation to the location of different feeding trees. Overall, Normand and Boesch (2009) suggested that based on their notions of distance and directions chimpanzees navigate across all parts of their range using a coordinate-base spatial representation.

In a study of decision-making, spatial mapping, and travel routes used by moustached tamarins (*Saguinus mystax*) and saddle-back tamarins (*Leontocebus fuscicollis*), Garber (1988, 1989, 2000) provides evidence that these primates navigate using a combination of route-based and coordinate-based systems of spatial representation. Data collected during a 3-month study in northeastern Peru indicated that 20 tree species accounted for approximately 75% of tamarins´ feeding time (Garber 1989). Tamarins commonly

Primate species	Summary of major findings	Reference
New World	**Monkeys**	
Alouatta palliata	-Goal-directed movement pattern. -Distance-efficient routes that interconnect feeding sites. -Use of a set of few "pivotal" trees.	Milton (1981, 2000)
Alouatta palliata	-Reuse of path segment during consecutive days. -Evidence of using a route-based map.	Jelinek et al. (2003), Garber and Jelinek (2006)
Alouatta palliata	-Reuse of pathways on different days.	Shaffer (2004)
Alouatta caraya	-Reuse of route quadrats through time.	Pereira (2004), Pereira et al. (2005)
Alouatta caraya	-The monkeys traveled using a non-random pattern.	Ventura (2004, 2005)
Alouatta caraya	-Spatial representation as a route-based map. -Use of routes interconnected with nodes.	Fernández (2008)
Ateles geoffroyi	-Goal-oriented travel pattern. -Use and reuse of arboreal pathways. -Traplining as a foraging strategy.	Milton (1981)
Ateles geoffroyi	-Use of a multiple central-place foraging strategy, that reduced the distance traveled between a given sleeping site and the feeding sites.	Chapman *et al.* (1989)
Ateles geoffroyi	-Use of a Lévy movement pattern. -Long travel paths for finding feeding sites and monitoring fruit status.	Ramos-Fernández *et al.* (2004)
Ateles geoffroyi	-Relatively straight-line travel between out-of-view feeding trees.	Valero and Byrne (2003, 2004, 2007)
Ateles belzebuth	-Travel along slopes. -Feeding trees nearby routes. Thus, potential monitoring of feeding tree status by using routes. -Reuse pathway segments interconnected by nodes (routes crossing points).	Suárez (2003)
Lagothrix lagotricha	-Route segments associated with topographic features of the home range. -Use of a limited set of route segments and nodes.	Di Fiore and Suárez (2004, 2007)
Pithecia pithecia	-Evidence of memory for locating, and knowing the productivity status of feeding trees. -Used feeding sites along daily pathways	Cunningham (2003), Cunningham and Janson (2003, 2007)
Saguinus mystax	-Use of quadrats that may function as switch points (landmarks). -Use of a route-based spatial representation, and eventually also a coordinate-based spatial representation as shorter distances to targets.	Garber (1989, 2000)
Leontocebus fuscicollis	*Ibid.*	Garber (1989, 2000)
Old World	**Monkeys**	
Papio haymadras	-Foraging strategy and navigation based on memory. -Knowledge of topography and consequent avoidance of potentially longer paths.	Altmann and Altmann (1970)
Papio haymadras	-Routes associated to a key goal: water holes. -Use of a single central sleeping place: A cliff.	Sigg and Stolba (1981)
Papio haymadras	-Evidence of using an "internal map" of their environment.	Kummer (1995)
Papio haymadras	-Spatial knowledge of large-scale home ranges, with the potential computational component when traveling from sleeping sites to targets. -During inter-group encounters, use of "detour path" to find targets. -Use of a route-based spatial representation.	Byrne (2000), Noser and Byrne (2004, 2007)
Macaca mulatta	-Use of "foraging routes" that interconnect different "roosting sites."	Makwana (1979)
Macaca fuscata	-Reuse of feeding sites by group members throughout different years.	Mori (2004a, 2004b: Pers. comm.)

Table 2.3. Examples of natural field studies and problem-oriented research on spatial memory in wild primates.

Primate species	Summary of major findings	Reference
Macaca fuscata	-Reuse of particular feeding trees through time. Some trees were selected more frequently than others. -Evidence of spatial knowledge of the distribution of feeding sites.	Nishikawa (2008)
Lophocebus albigena	-When the monkeys move to a new area, initially the movement between feeding sites was no-distance efficient. Later, with their own environment familiarity, travel distance decreased. -Primates "accumulative" of "spatial memories."	Janmaat *et al.* (2008)
Apes		
Pongo pygmaeus	-Use of efficient routes for reaching feeding sites located at long distances. -Possibility of monitoring fruit maturation and productivity status.	Galdikas and Vasey (1992)
Pan troglodytes	-Distribution of hammers and their transportation resembles a coordinate-based spatial representation. -Evidence of mental configuration about the direction to reach the goals. Goals are reached from multiple directions. -Evidence of "permutation" while transporting hammers to particular feeding sites, possibly forming a coordinate-based map. -Low circuity index between feeding sites (1.04).	Boesch and Boesch (1984), Normand and Boesch (2009)
Prosimians		
Microcebus murinus	-Avoid using random movements, possibly used a topological representation. -Use by primates by high-efficiency travel segments to reach plat-forms; thus indicating that some other type of spatial representation may be in play.	Lührs *et al.* (2007, 2008)
Eulemur fulvus rufus	-Possibly use a route-based map. – Primates mainly travel to potential nearby feeding sites taking into consideration inter-patch distances and turning angles.	Erhart and Overdolff (2008)
Propithecus edwardsi	*Ibid.*	Erhart and Overdolff (2008)

Table 2.3. continued.

visited several trees of the same species each day and did so using relatively straight-line progression. On average the distance between sequential feeding sites was ~150 m. Tamarins reached these trees from a variety of different directions and rarely backtracked. Garber (1989) concluded that the tamarins recall the spatial position of a large number of trees in their home range and base foraging decisions based on spatial memory.

In a related study of nectar-feeding of moustached and saddle-back tamarins (Garber 1988), there was evidence that both distance and the size of food reward were factors used by tamarins in making foraging decisions. On 86% of occasions, tamarins were found to select the nearest available flowering tree as their next feeding site. On those occasions in which the nearest tree was not selected, the second nearest tree was selected. Using this dataset, Garber (2000) tested Poucet's model of spatial representation in large and small-scale space. He found that the most commonly used feeding sites were located in 6 quadrats (50 × 50 m) that were scattered across the group´s range. Tamarins repeatedly reused a set of route segments between 150–350 m and traveled in relatively straight-line between these areas. Across the group´s 40-ha home range, Garber (2000) identified 18 areas that functioned as landmarks or switching points. From these nodes, the tamarins change directions of travel. Garber (2000) also found that once the tamarins arrived at 50–100 m of the feeding sites, they traveled directly to the feeding sites. Tamarin travel patterns are consistent with Poucet´s model of spatial representation suggesting the use of a route-based spatial representation in large-scale space, and a coordinate-based spatial representation in small-scale space.

Thus considering the previous information, the degree to which other primates rely on landmarks, compute route segments and use spatial information differently remains unclear. However, given that the ability to navigate throughout the space not only by using a route-based spatial representation but also by using a coordinate-based spatial representation appears to be present in monkeys and apes, it may be a characteristic of most primates.

Examining Spatial Representations in Capuchins

Capuchins have been the subjects of many experimental studies designed to test their cognitive capabilities (see overview in Fragaszy *et al.* 2004). This section reviews research focusing on studies of capuchin spatial representation. For example, in a series of **captive studies**, De Lillo *et al.* (1997) presented brown capuchins (*Sapajus apella*) with the task of locating food rewards concealed in 9 containers distributed in different spatial configurations within a small captive enclosure measuring 3 × 3 x 2.5 m. All containers were visible to the capuchins but the food was hidden. In the first set of experiments, a baseline condition of a 3 × 3 matrix (9 containers at 70 cm between each container) was presented. This was followed by an arrangement in which a triangle configuration was created using three clusters of containers, each cluster separated by distances of 1.1 m. The clusters were formed by three containers each located 35 cm from the other. In the final trial, the original baseline configuration was presented to the capuchins (De Lillo *et al.* 1997). In the second experiment, the capuchins visited all containers in a cluster before visiting the next groups of containers (50 out of 60 trails). De Lillo *et al.* (1997) suggested that capuchins followed a strategy in which nearer items are visited in sequence, and hypothesized that wild capuchins may use a similar searching schema to exploit patchily distributed food sources.

In a second set of tests, the experimental arrangement included food distributed in a 3 × 3 matrix (60 × 60 cm between containers), a circle (60 cm between containers, and 1.2 m radius), a straight line (40 cm between containers), and an X-shaped line configuration (60 cm between containers, and 2.1 m between containers located at the vertex) (De Lillo *et al.* 1998). Based on 540 trials, the results indicated that the capuchins searched containers in a sequence that limited backtracking or revisits to previously exploited containers. The authors concluded that the capuchins remembered which of the nine containers they had previously visited, and used this information to avoid revisiting depleted containers (De Lillo *et al.* 1998).

De Lillo *et al.* (1997, 1998) found that when resources were arranged such that the food resources were presented in clumped units, the capuchins exploited containers that were in closer spatial proximity (e.g. De Lillo *et al.* 1997: 35 cm *vs.* 1.1 m) before visiting other containers. Following these results, the authors argued that in the wild, capuchins may "view" a set of nearby resources as a single foraging patch (from field research, see: Robinson 1986, Chapman 1988, Di Bitetti 2001), and therefore, chunk or group different food patches as a single spatial unit.

In a second set of captive experiments in small-scale space, brown capuchins (*S. apella*) were allowed to observe the rotation of a platform (36 cm in diameter) containing a concealed food reward (a box) and an associated visual cue (black cylinder) (Potì 2000). During the experiment, the box with the reward was rotated out of view; however, the landmark cue remained visible. In the first experiment the reward box (5.5 × 6.5 cm) was located perpendicular (90°, 180°, 270°, and 360°) and at a distance of 9 cm distance from the landmark. A second unrewarded box without a nearby landmark was located at a parallel position to the rewarded box on the opposite side of the experimental area (no angle and distance provided). The experimental conditions included phases in which the boxes were rotated in view and out of view of the capuchins. Potì (2000) found that the capuchins successfully used the spatial relationship between the platform and the landmark cue to locate the food reward after observing the rotation. However, when the capuchins were unable to observe the rotation process, they tended to rely on the position of one side of their body relative to the platform (egocentric information) to locate the food reward. The capuchins were successful 94–97% of the time in the control condition and 64–67% in the test condition (Experiment 1), 87.5–100% during the visible rotation condition, and 99–100% during the invisible rotation condition in Experiment 2. Thus, Potì (2000) concluded from these experiments that in small-scale space brown capuchins used both egocentric and geocentric frames of reference to locate food rewards.

Potì *et al.* (2004, 2005) tested the ability of capuchins to use visual cues such as red plastic cylinders to locate food rewards. In the first experiment, the monkeys were tested using four landmarks arranged within a configuration of 11 × 11 containers filled with dark water in a small-scale space (area of 3 × 3.5 × 2.6 m). To solve the problem the capuchins were required to search for the reward within the square formed by the four landmarks. In a second experiment, the capuchins were presented with only two landmarks and a food reward located between the cues. In the first experiment, the capuchins preferentially searched for rewarded containers near each landmark rather than searching between landmark configurations. In the second experiment, capuchins used the closest single landmark to reach the containers. Thus, the monkeys did not rely on the landmark configuration. The authors suggested that under the conditions of these experiments, the capuchins encoded the location of food rewards in relation to single visual cues. This was a less efficient spatial strategy and suggested that capuchins did not rely on landmark configurations in small-scale space.

Using a two-dimensional alley maze displayed on a computer screen (46 × 28 cm), Fragaszy *et al.* (2003) tested the ability of capuchins to plan sequential movement. Three capuchins (*S. apella*) were presented with multiple non-symmetrical mazes containing starting and endpoints. After 192 pre-test trials, the capuchins were tested to determine their frequencies of error and error-free maze trials. Two of the capuchins performed fewer errors than expected by chance (Fragaszy *et al.* 2003). Based on these results, the authors suggested that the capuchins were able to effectively plan a route by which to navigate the maze.

Experimental field studies of decision-making and spatial memory have been conducted on several species of Neotropical primates including tamarins, night monkeys, and capuchins (e.g. Garber and Dolins 1996, Janson 1996, 1998, 2007; Janson and Di Bitetti 1997; Garber and Paciulli 1997, Bicca-Marques 1999, Bicca-Marques and Garber 2003, Garber and Brown 2006, Urbani this study). Janson and colleagues conducted a series of innovative field experiments on spatial memory in *Sapajus nigritus* in Argentina (Janson 1996, 1998, 2007; Janson and Di Bitetti 1997; Di Bitteti and Janson 2001). Their experiments used 17 feeding platforms of different sizes distributed across an area of approximately 800 × 500 m. Feeding platforms contained different quantities of food rewards. The study tested the ability of wild capuchins to evaluate tradeoffs in distance traveled to reach feeding sites *vs.* quantity in selecting feeding sites, as well as strategies of food detection in locating and monopolizing feeding sites. Capuchins preferentially selected more productive over less productive feeding sites (as low as four *vs.* as many as 16 tangerines), even when doing so resulted in increased travel distance (Janson 1996, 1998; Janson and Di Bitteti 1997). These results indicate that capuchins encoded the location of a feeding site with expectations of food quantity, and integrated this information into a foraging strategy (Janson and Di Bitteti 1997). A similar ability was found in wild moustached tamarins (*Saguinus mystax*) and saddleback tamarins (*Leontocebus fuscicollis*) (Garber 1988).

Janson (2007) located three platforms in an oblique triangle configuration in the home range of a *S. nigritus* group. The distance between platforms ranged from 294 m to 665 m. The amount of food rewards on platforms varied from five to up to 80 pieces of fruit (tangerines and bananas). A platform visit sequence began once the monkeys had visited at least two platforms. Based on 136 platform visits over the course of four years, the results indicate that capuchins were found to select nearer feeding sites even when they contained lower food reward (Janson 2007).

Garber and associates examined decision-making, foraging strategies, and use of landmark cues in a group of wild Costa Rican white-faced capuchin monkeys (*Cebus imitator*) in small-scale space (Garber 2000, Garber and Paciulli 1997, Garber and Brown 2006). In one field experiment testing the capuchins' ability to use spatial and quantity information, individuals initially were found to associate spatial information with the presence/absence of concealed food rewards (five of 13 experimental feeding platforms contained food rewards) but not with the size of the food rewards (½ banana *vs.* three bananas). After a period of five days, the capuchins preferentially selected platforms with a higher quantity of concealed food rewards over feeding platforms with a lower quantity of concealed food rewards. Thus, in terms of a hierarchy of cues, the capuchins first learned presence *vs.* absence of food rewards and secondly higher *vs.* lower food rewards (Garber and Paciulli 1997).

In another set of experiments, Garber and Brown (2006) tested the ability of wild white-faced capuchins to predict the spatial location of provisioned feeding sites using a set of two or three landmark cues to locate hidden food rewards. The experimental design involved a feeding station (8 m diameter) composed of eight identical platforms arranged in a circle. The only information available to the forager to identify which two of eight platforms contained a food reward was a series of a 2 m high pink-yellow poles. Across a series of experiments, the poles were located at 2 m distance from the reward platforms. Although which two of the eight platforms contained a food reward (two bananas) was random, the spatial relationship between the poles and the reward platform was constant. A total of six experimental configurations were presented with two, three, and no landmarks. In order to solve the foraging task the capuchins were required to use the landmarks as an array (not as single landmarks). The results indicate that the capuchins successfully encoded the spatial relationship between two and three landmark cues to predict the location of food rewards in small-scale space. In addition, based on the sequence of platform selection there is evidence that the capuchins were capable of mentally rotating landmark arrays to locate baited feeding sites.

Given the evidence and reports of the ability of wild capuchins to reuse the same set of travel paths when traveling in their home ranges (*S. nigritus*: Kühlorn 1939, Janson 1996; *Cebus olivaceus*: Robinson 1986; *C. imitator*: Urbani 2004), during 2006, I conducted a *natural field study* to test hypotheses to determine how white-faced capuchins represent salient features in their environment in large- and small-scale space. The results are presented in this monograph.

In sum, there is evidence that in captivity, capuchin spatial memory is characterized by: (a) reliance on both egocentric and geocentric information to locate food rewards, (b) use of single landmarks in forming spatial relationships with reward sites in small-scale space, (c) visiting food patches that reduce backtracking and revisiting to depleted feeding sites, and (d) evidence that capuchins may plan routes when navigating in small-scale space. In wild capuchins in small-scale space, there is: (a) evidence that two and/or three landmarks can be used as an array to locate feeding sites, and (b) evidence that capuchins first encode spatial information in terms of the presence or absence of a food reward, and then quantity information concerning differences between lower or higher food rewards. In large-scale space, wild capuchins had been found to select the nearest feeding sites over feeding sites located at longer distances. However, considering the previous research, there is still limited research about how capuchins´ spatial skills vary when traveling between targets and at the vicinity of those targets.

The goal of this study is to examine: (a) the ability of wild capuchins to locate and relocate feeding sites, (b) their selection of travel routes, and (c) the manner in which landmarks are used as nodes to re-orient travel. Finally, in order to evaluate the theoretical problems related to the manner in which non-human primates integrate spatial information during navigation, I test hypotheses concerning the ability of capuchins to use route-based spatial representation and/or coordinate-based spatial representation in large- and small-scale space.

3
_____

Study Site, Study Group, Materials, and Methods

Study Site

This field study was carried out in the private forest reserve of La Suerte Biological Field Station (Estación Biológica La Suerte, EBLS) located at 75 m asl., ~12 km west of the town of Cariari, La Rita District, Pococí County, Limón Province, northeastern Costa Rica (10°26'N; 83°47'W; Figure 3.1). A fixed point located at UTM: 1.155.195N; 194.930E (WGS 1984 Zone 16N) was used to calibrate all GPS UTM points recorded in the field. The fixed point was the rain gauge placed near the main building of the field station (see the first map of the Appendix).

The region of Cariari, and the EBLS, are part of the "Tortuguero Conservation Area – Área de Conservación Tortuguero: ACTO-" (SINAE 2001). La Suerte Biological Field Station is 283.3 ha (700 acres) in area (EBLS 2009). The forest fragment used for the present study is locally referred to as the "small forest" (24.5 ha). The field station has been used for ecological studies, training, and education by researchers, students, and members of the local community since 1994. A review of published research conducted at La Suerte Biological Field Station is found in Bolt *et al.* (2021). Table 3.1 summarizes the time period and activities for the present study.

The vegetation of the Tortuguero Conservation Area is classified as "Bosque muy húmedo tropical, Very humid tropical forest; bmh-T" (Bolaños *et al.* 1993). The study site represents one of the few remaining primary Atlantic lowland rainforests in this part of the country (Aguilar 2007: Pers. comm.). In order to determine the botanical composition of the "small forest," a total of 20 transects of 50 m x 2 m in size were surveyed by measuring every tree with a diameter at the breast (DHB) ≥10 cm according to A. Gentry's Forest Transect Method (Oliver and Miller 2002). This work was conducted with the cooperation of the Costa Rican botanist Reinaldo Aguilar. Although the area sample was relatively small (2000 m²), the purpose of these transects was to identify the floristic composition of the "small forest" which included most of the home range of the capuchin study group. I used this sample to determine whether the capuchin's selection of feeding/resting trees within the "small forest" was random or targeted to particular species. The average DBH of trees in the "small forest" was 45.1±45.8 cm (*n*=157). Sixty-three percent of the trees samples were less than 50 cm DBH. A total of 75 tree species (Table 3.2), and 37 tree families

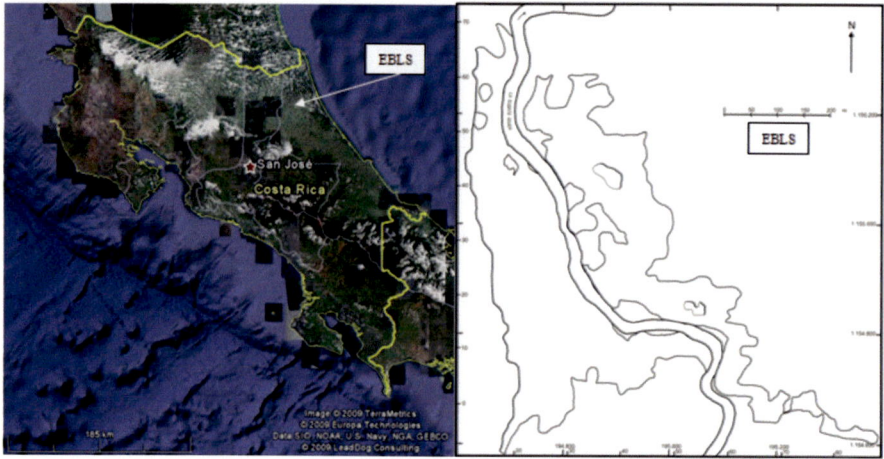

Figure 3.1. Map of the location of La Suerte Biological Field Station and its "Small Forest," northeastern Costa Rica (Images: Google Earth™, January 2009); ArcGIS generated map of the "Small Forest" limits (this study, see Appendix). *Note:* The arrow indicates the approximate location of the EBLS.

Activity	Start Date	End Date	Consecutive Days (Observation hours)
Grid system placement + vegetation characterization + pre-observation period	24/January/2006	18/February/ 2006	26 days (approx. 644 three persons/hours)
Follow 1 + botanical study	1/March/2006	14/March/2006	12 days + two botany days
Follow 2 + botanical study	1/April/2006	14/April/2006	12 days + two botany days
Follow 3 + botanical study	1/May/2006	14/May/2006	12 days + two botany days
Follow 4 + botanical study	3/June/2006	16/June/2006	12 days + two botany days
Follow 5 + botanical study	1/July/2006	14/July/2006	12 days + two botany days
Follow 6 + botanical study	1/August/2006	14/August/2006	12 days + two botany days
Platforms placement for Experiment # 1	31/August/2006	31/August/2006	1 day. Note: This day is also a pre-bait day (see below).
Pre-bait of platforms for Experiment # 1	31/August/2006	4/September/ 2006	5 days
Experiment # 1 + Follow 7 + botanical study	5/September/2006	28/September/ 2006	24 days + two botany days Note: Apart from the experimental study, the behavioral-ecological data were collected during these days as in a regular "Follow."
Platforms placement for Experiment # 2	29/September/2006	29/September/ 2006	1 day (approx. 36 four persons/hours) Note: This day is also considered a pre-bait day (see below).
Pre-bait of platforms for Experiment # 2	29/September/2006	10/October/ 2006 (see below)	12 days. Note: The baited platforms were not visited until 11/October/2006. On 30/September/2006 the capuchins moved out of EBLS property, so the part of the forest with the platforms was not visited this day.

Table 3.1. Field research schedule and research-time investment between January 2006 and April 2007.

Activity	Start Date	End Date	Consecutive Days (Observation hours)
Follow 8 + botanical study	1/October/2006	10/October/ 2006	10 days + two botany days: see below Note: The platforms were baited, but never visited during a 10-day period. Therefore, this period was considered not only a pre-baited period but also a regular behavioral-ecological "Follow."
Experiment # 2 + Follow 9 + botanical study	11/October/2006	25/November/ 2006	43 days + two botany days Note: Apart from the experimental study, the behavioral-ecological data was collected during these days as in a regular behavioral-ecological "Follow."
Follow 10 + botanical study	1/December/2006	14/December/ 2006	12 days + two botany days
Botanical study	22/April/2007	25/April/2007	4 days (approx. 93 three persons/hours)
			230 total fieldwork days

Table 3.1. continued.

Species	Tree Family	# Individuals	% Individuals	Average DBH	Estimated Ind./Ha
Pentaclethra macroloba	Fabaceae/Mimosaceae	27	17.2	71.7	135
Goethalsia meiantha	Tiliaceae	16	10.2	53.6	80
Calatola costarricensis	Icacinaceae	5	3.2	35.5	25
Alseis costaricensis	Rubiaceae	4	2.6	89.0	20
Hieronyma alchorneoides	Euphorbiaceae	4	2.6	50.4	20
Socratea exorrhiza	Arecaceae	4	2.6	17.4	20
Byrsonima crispa	Malpighiaceae	3	1.9	45.5	15
Casearia arborea	Flacourtiaceae	3	1.9	19.3	15
Mortoniodendron spp.	Tiliaceae	3	1.9	51.2	15
Ocotea cernua	Lauraceae	3	1.9	23.0	15
Apeiba membranacea	Tiliaceae	2	1.3	83.0	10
Bravaisia integerrima	Acanthaceae	2	1.3	20.8	10
Brosimum alicastrum	Moraceae	2	1.3	169.0	10
Capparis pittieri	Capparaceae	2	1.3	12.5	10
Celtis schippii	Ulmaceae	2	1.3	20.0	10
Chrysophillum spp.	Sapotaceae	2	1.3	16.3	10
Cordia bicolor	Boraginaceae	2	1.3	22.0	10
Croton schiedeanus	Euphorbiaceae	2	1.3	15.3	10
Dendropanax spp.	Araliaceae	2	1.3	33.5	10
Dipteryx panamensis	Fabaceae/Papilonaceae	2	1.3	250.0	10 (0.5)
Guarea la selva	Meliaceae	2	1.3	18.8	10
Guatteria aeruginosa	Annonaceae	2	1.3	22.8	10
Hampea appendiculata	Malvaceae	2	1.3	36.0	10
Inga sp.	Fabaceae/Mimosaceae	2	1.3	34.5	10
Protium spp.	Burseraceae	2	1.3	29.5	10
Sapium grandulosum	Euphorbiaceae	2	1.3	64.3	10
Simarouba amara	Simaroubaceae	2	1.3	55.0	10
Simira maxonii	Rubiaceae	2	1.3	34.8	10
Tabebuia spp.	Bignoniaceae	2	1.3	46.5	10
Virola koshnyi	Myristicaceae	2	1.3	29.5	10
Virola sebifora	Myristicaceae	2	1.3	37.5	10

Table 3.2. Floristic composition by tree species of La Suerte Biological Field Station "Small Forest."

Species	Tree Family	# Individuals	% Individuals	Average DBH	Estimated Ind./Ha
Ampelosera macrocarpa	Ulmaceae	1	0.64	12.0	5
Anaxagorea phaeocarpa	Annonaceae	1	0.64	11.0	5
Cedrela spp.	Meliaceae	1	0.64	77.0	5
Ceiba pentandra	Bombacaceae	1	0.64	190.0	5
Cordia cymosa	Boraginaceae	1	0.64	19.0	5
Cordia lucidula	Boraginaceae	1	0.64	10.5	5
Cordia spp.	Boraginaceae	1	0.64	31.0	5
Coussarea spp	Rubiaceae	1	0.64	11.5	5
Culubrina spinosa	Rhamnaceae	1	0.64	13.5	5
Cupania pseudoestipularis	Sapindaceae	1	0.64	23.5	5
Cynometra retusa	Fabaceae/Caesalpinaceae	1	0.64	65.0	5
Duguetia panamense	Annonaceae	1	0.64	21.0	5
Faramea occidentalis	Rubiaceae	1	0.64	22.0	5
Ficus spp.	Moraceae	1	0.64	110.	5
Guarea aff *bullata*	Meliaceae	1	0.64	17.5	5
Guateria auriginosa	Annonaceae	1	0.64	31.0	5
Hasseltia spp.	Flacourtiaceae	1	0.64	19.5	5
Hernandia didymantha	Hernandiaceae	1	0.64	17.5	5
Hymenolobium mesoamericanum	Fabaceae/Papilonaceae	1	0.64	86.0	5
Ilex skutchii	Aquifoliaceae	1	0.64	46.0	5
Inga ciliata	Fabaceae/Mimosaceae	1	0.64	12.0	5
Lacistema aggregatum	Flacourtiaceae	1	0.64	14.0	5
Laetia procera	Flacourtiaceae	1	0.64	48.5	5
Lecythis ampla	Lecythidaceae	1	0.64	95.0	5
Lonchocarpus ferrugineus	Fabaceae/Papilonaceae	1	0.64	69.0	5
Lonchocarpus pentaphyllus	Fabaceae/Papilonaceae	1	0.64	18.0	5
Minquartia guianensis	Olacaceae	1	0.64	22.0	5
Myriocarpa longipes	Urticaceae	1	0.64	10.0	5
Nauclopsis spp.	Moraceae	1	0.64	18.0	5
Ochroma pyramidale	Bombaceae	1	0.64	102.	5
Otoba novogranatensis	Myristicaceae	1	0.64	15.5	5
Poulsenia armata	Moraceae	1	0.64	173.0	5
Pouteria aff *viridis*	Sapotaceae	1	0.64	16.5	5
Pouteria silvestris	Sapotaceae	1	0.64	73.5	5
Pouteria torta	Sapotaceae	1	0.64	19.5	5
Quararibea bracteolosa	Bombaceae	1	0.64	12.5	5
Quararibea obliquifolia	Bombacaceae	1	0.64	20.0	5
Sapranthus viridiflorus	Annonaceae	1	0.64	28.0	5
Stemmadenia donnell-smithii	Apocynaceae	1	0.64	22.0	5
Stephanopodium costaricense	Fabaceae/Mimosaceae	1	0.64	42.0	5
Terminalia oblonga	Combretaceae	1	0.64	138.5	5
Theobroma spp.	Sterculiaceae	1	0.64	17.0	5
Unonopsis pittieri	Annonaceae	1	0.64	11.0	5
Total = 73	37	157	100.	45.1	

Note: The data in parenthesis for the density of *Dipteryx panamensis* is based on a total survey of this species in the 23.7 ha of the "Small Forest."

Table 3.2. continued.

Family	# Individuals	% Individuals	Average DBH
Fabaceae/Mimosaceae	30	19.1	70.7
Tiliaceae	21	13.4	56.0
Euphorbiaceae	9	5.7	38.1
Annonaceae	8	5.1	21.5
Rubiaceae	8	5.1	57.4
Flacourtiaceae	6	3.8	23.3
Boraginaceae	5	3.2	20.9
Fabaceae/Papilonaceae	5	3.2	134.6
Icacinaceae	5	3.2	35.5
Moraceae	5	3.2	127.8
Myristicaceae	5	3.2	29.9
Sapotaceae	5	3.2	28.4
Arecaceae	4	2.6	17.4
Lauraceae	3	1.9	23.0
Malpighiaceae	3	1.9	45.5
Meliaceae	3	1.9	35.8
Ulmaceae	3	1.9	17.3
Acanthaceae	2	1.3	20.8
Araliaceae	2	1.3	33.5
Bignoniaceae	2	1.3	46.5
Bombacaceae	2	1.3	105.0
Bombaceae	2	1.3	57.3
Burseraceae	2	1.3	29.5
Capparaceae	2	1.3	12.5
Malvaceae	2	1.3	36.0
Simaroubaceae	2	1.3	55.0
Apocynaceae	1	0.64	22.0
Aquifoliaceae	1	0.64	46.0
Combretaceae	1	0.64	138.5
Fabaceae/Caesalpinaceae	1	0.64	65.0
Hernandiaceae	1	0.64	17.5
Lecythidaceae	1	0.64	95.0
Olacaceae	1	0.64	22.0
Rhamnaceae	1	0.64	13.5
Sapindaceae	1	0.64	23.5
Sterculiaceae	1	0.64	17.0
Urticaceae	1	0.64	10.0
Total = 37	157	100.0	45.1

Table 3.3. Floristic composition by tree families of La Suerte Biological Field Station "Small Forest."

(Table 3.3) were recorded. The most common tree species was *Pentaclethra macroloba* with a density of 150 individuals/ha, followed by *Goethalsia meiantha* with 59 individuals/ha. The sample also includes *Calatola costarricensis, Alseis costaricensis, Hieronyma alchorneoides, Socratea exorrhiza, Casearia arborea, Byrsonima crispa, Mortoniodendron* sp., and *Ocotea cernua*. The families that dominated the forest are Fabaceae/Mimosaceae, Tiliaceae, and Euphorbiaceae.

A total of 10 vegetation categories were identified within the total forested area of 24.5 hectares used by the capuchin study group at the EBLS. First, during the placement

Type of vegetation	Abbreviation	Ha	%	Brief definition
Advanced secondary forest	ASF	6.7	27.0	A tall regenerated forest composed with scattered old trees. This may have Caribe pines (Pinus caribaea; ASF-PI) or pejivaye (Bactris gasipaes; ASF-PE).
Bambuzal	B	0.64	2.6	It refers to the presence of naturally grown and highly tangled bamboo.
Cañal	C	0.13	0.55	It implies the location of naturally grown cane plants.
Charral	Ch	1.2	5.1	It implies de presence of fern and/or "zacate." They form gaps within the forests.
Gap	G	0.28	1.1	There are two types of gaps that are defined depending on how they were formed. As follow, by tree fall (G-TF) or by riverine flows (G-R).
Platanillal	Pl	0.30	1.3	It is a conglomerate of platanillo (Heliconia spp.) plants. Their height is between 2 and 3 m. They form gaps within the forest.
Primary forest	PF	8.6	35.3	Dense, old, and tall original forest.
Secondary forest	SF	1.1	4.4	A forest in process of regeneration with trees of circa 6–10 m with at least 12 years old.
Swamp	Sw	0.16	0.65	Any water pod within the study area, normally with aquatic plants.
Tacotal or shrub.	T	5.4	22.0	It is formed by low and tangled vegetation, sometimes with "palmito" or young pejibaye (T-P). They form gaps within the forest.
Total	-	24.5	100	

Note: Around the "Small Forest" there are other vegetation categories: Grasses, plantations/crops, living fences, and pastures.

Table 3.4. Vegetation categories at La Suerte Biological Field Station "Small Forest."

of the grid system (see Table 3.1), a local field assistant, a field topographer (+40 years of experience in the Neotropics, F. Urbani) and the author (B. U.) walked each part of the "small forest." At this time, an initial classification of vegetation areas was made directly in the forest and represented on the field map. Between March and May 2006, three different field assistants assisted in the characterization of the forest by walking the trails.

In 2007, an experienced Costa Rican botanist (Reinaldo Aguilar), a local field assistant, and the author (B.U.) compiled a final list of vegetation types by walking the complete "small forest," and conducting an examination of the floristic composition of the area. The field map was digitized with the use of ArcGIS®. These maps show the vegetation matrix of the area (see Appendix). Table 3.4 defines each vegetation category and presents the total area for each vegetation type. Advanced secondary forest and primary forest are the most common types of vegetation in the capuchins´ home range.

Rainfall and temperature at the field station were recorded on a daily basis (~5:15 pm) with a digital thermometer and a rain gauge (circa 14 days per month). Based on 158 days of data collection, the annual temperature averaged 28.3°C (maximum: 29.1°C in November, and minimum: 27°C in December; Figure 3.2). Annual rainfall, also based on 158 days of data collection, was 3,214 mm (maximum: 653 mm in December and minimum: 103 mm in May; Figure 3.3). The two greatest peak periods of rainfall occurred during the first 4 days of December (mean=261 mm) and on July 13th, 2006 (123 mm), both the result of tropical rainstorms.

The topsoil humidity of the EBLS "small forest" was determined by taking three random samples of the soil at a depth of 10 cm, and six samples, one in each of six different soil types in a recently-opened trench (~2 m in deep) located in a pasture at the border of the

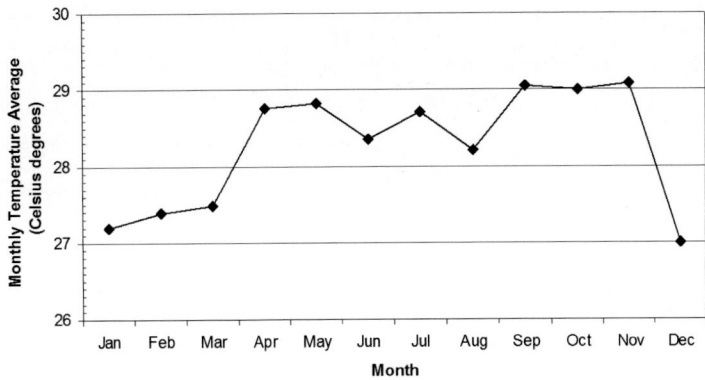

Figure 3.2. Temperature at La Suerte Biological Field Station between January and December 2006.

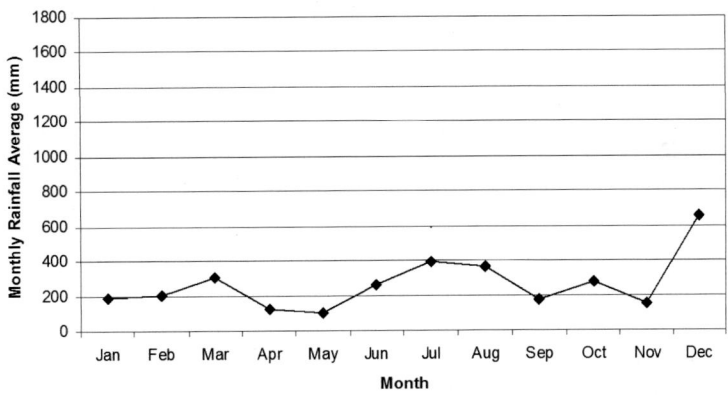

Figure 3.3. Rainfall at La Suerte Biological Field Station between January and December 2006.

EBLS "small forest." The mean average soil humidity of these samples was 27.7±11.0% (n=9). These data are similar to those reported at La Selva Biological Station also located in the northern Atlantic region of Costa Rica, approximately 50 km northwest of the EBLS. Soils with high humidity contain macro-pores that absorb water rapidly (Sollins *et al.* 1994).

Two random soil samples (from the forest floor and the trench) were analyzed by X-ray diffraction to determine mineralogical composition. The results indicate that clay minerals (65%) are the most prevalent in the soil, followed by quartz (23.5%), feldspar (10.5%), and siderite (1%). The separated clay fraction (<2 µm) is formed by chlorite (17.5%), illite (31%), illite-smectite (44%), and kaolinite (7.5%). This soil composition also is similar to that of La Selva Biological Station and is typical of Central American rainforests (Sollins *et al.* 1994).

Table 3.5 shows the water composition at the EBLS collected from rainfall, the La Suerte River, and an inner forest creek and pond. Water at the EBLS has a pH that ranges from 5.5 (rainfall sample) to 7.7 (La Suerte River sample), with an average of 6.6±0.9. The geological formation at the EBLS is consistent with Quaternary sediments and Quaternary volcanic rock from areas characterized by volcanic activity (Tournon and Alvarado 1995).

Water sample ID (Collection location)	Si	Ca	Mg	Na	K	Fe	Mn	F	Cl	SO$_4$	CO$_3$	HCO$_3$	pH	Conductivity
La Suerte River (grid point 32/20)	21	14	<1	4.8	2.7	<1	<1	<1	3.5	5	<10	81	7.7	117
Inner forest water pond	3.6	2.5	<1	<1	0.5	<1	<1	<1	2.4	1.4	<10	20	6.4	32
Inner forest creek (grid point 29/6)	9.7	5.4	<1	2.8	1.2	<1	<1	<1	2.6	<1	<10	<10	6.9	56
Rainfall water	<1	<1	<1	1.1	<1	<1	<1	<1	5.1	1.4	<10	39	5.5	23
Average	8.8	5.7	<1	2.4	1.4	<1	<1	<1	3.4	2.2	<10	38	6.6	57
SD ±	8.9	5.8	0	1.8	1	0	0	0	1.2	1.9	0	31	0.9	42
Min.	<1	<1	<1	<1	<1	<1	<1	<1	2.4	<1	<10	<10	5.5	23
Max.	21	14	<1	4.8	2.7	<1	<1	<1	5.1	5	<10	81	7.7	117

Note: Chemical components expressed in ppm. Conductivity in µS/cm.

Table 3.5. Water analysis composition at La Suerte Biological Field Station, 2006.

Study Group

The capuchin study group was named *La Yunai*. It is a fully-habituated troop of white-faced capuchin monkeys (*Cebus imitator*) that range across the "small forest" at EBLS and surrounding properties. This group has been the focus of several field studies and is the only group that permanently inhabits this forest fragment (Garber and Paciulli 1997, Garber and Rehg 1999, Garber and Brown 2004, 2006, Urbani 2004, 2009, Bezanson 2006, Garber *et al.* 2008). Group members were identified based on physical characteristics including body size, facial markings, hair patterns and coloration, presence of dependent offspring on females, and in some cases feet/shoulder scars (see: Table 3.6). The group size varied from 11 to 15 individuals (mean=12±1.3) based on 29 censuses between March and December 2006 in which all individuals were visible at the same time (Table 3.7). Variation in group counts was due to the disappearance of young individuals. This group size is similar to that reported for the same capuchin troop (12–14 individuals) in a long-term study conducted at the EBLS between 2002 and 2003 (Bezanson 2006). White-faced capuchins have been reported to live in groups that range in mean size from 12 to 33 members.

The La Yunai group was composed, by the end of the field study (December 2006), of two adult males, one subadult male, four adult females, one subadult female, one male juvenile, and two juveniles of unknown sex. The age/sex composition of the La Yunai group is similar to that reported for other groups of *Cebus imitator* and other capuchin species (Tables 3.8, 3.9; Fragaszy *et al.* 2004). The adult male:adult female ratio (0.74), immature:adult female ratio (1.4), and the infant:adult female ratio (0.38) of the La Yunai group also were similar to the averages reported for *Cebus imitator* by Fragaszy *et al.* (2004). Two other primate species are present at the EBLS, mantled howler monkeys (*Alouatta palliata*) and the Central American spider monkeys (*Ateles geoffroyi*; a solitary individual sporadically visited the "small forest").

Table 3.6 (opposite above). Members of La Yunai group at La Suerte Biological Field Station between March-December 2006.

Table 3.7 (opposite below). Demography and composition of La Yunai group at La Suerte Biological Field Station between March-December 2006.

Individual name	Sex/Age Class	Comments
Ramón	α ♂ adult	Also known as "Rorro" during this field study
Ñoño	β ♂ adult	-
Jirafales	♂ Subadult	-
Florinda	α ♀ adult	-
Chilindrina	β ♀ adult	-
Fuzzy	♀ adult	-
Bruja	♀ adult	-
Patty	♀ Subadult	-
Glorio/a	Unknown sex juvenile	Mother: Unknown. Born: Unknown, but pre-March 2006. Dead: August 2006.
Popis	Unknown sex juvenile	Mother: Unknown. Born: Unknown, but pre-March 2006. Dead: June 2006.
Barrigo/a	Unknown sex juvenile	Mother: Unknown. Born: Unknown, but pre-March 2006.
Kiko	♂ juvenile (infant)	Presumed mother: Florinda (Kiko was carried on Florinda's back by March 2006). Born: Unknown, but pre-March 2006. Considered an infant between March-May 2006 because Kiko presented no independence in locomotion during these dates. At the time of it first observation on March 1st, 2006, Kiko was younger than Chavo/a. Thus, Kiko was probably born during the last days of February 2006.
Chavo/a	Unknown sex juvenile (infant)	Mother: Chilindrina. Born: April 2006. Considered infant between April-June 2006 because this monkey presents no independence in locomotion during these dates.
Jaimito/a	Unknown sex infant	Mother: Patty. Born: 3/July/2006, Disappeared: August 2006.
Godino/az	Unknown sex infant	Mother: Fuzzy. Born: May 2006, Disappeared: June 2006.
Setenta-y-uno/a	Unknown sex infant	Mother: Bruja. Born: May 2006, Disappeared: June 2006.

Definitions: Infants = individuals that were able to survive after birth, and weaning from their mother for a limited period of time (~3 months at La Suerte). Infants travel on mother's dorsal area. The body color is dark grayish. Juveniles = young individuals which perform all their activities independently and who no longer nurse from their mothers. At La Suerte, capuchins are totally independent at the age of four months, (Urbani 2006 Pers. obs.; Bezanson 2006: Pers. comm.). This differs from other sites in Costa Rica such as Santa Rosa, at which capuchins are independent at age 7 months. I only saw two bouts of adult females nursing juveniles (see Table 4.1). Subadults = individuals of approximately three to five years of age, larger than juveniles but still smaller than fully developed adults. The male subadult size is similar to that of an adult female. Subadults are designated as "pre-adult individuals" at La Suerte by Bezanson (2006), "large juveniles" by MacKinnon (2002), and young adults by our observation team at La Suerte. Adults = sexually mature individuals (>5 years), full size depending on their sex; males are larger than females. In some females, nipples are visible, and in males, the testicles are sometimes conspicuous.

Individual name	Mar	Apr	May	Jun	Jul	Aug	Sep	Oct	Nov	Dec
Ramón	X	X	X	X	X	X	X	X	X	X
Ñoño	X	X	X	X	X	X	X	X	X	X
Jirafales	X	X	X	X	X	X	X	X	X	X
Florinda	X	X	X	X	X	X	X	X	X	X
Chilindrina	X	X	X	X	X	X	X	X	X	X
Fuzzy	X	X	X	X	X	X	X	X	X	X
Bruja	X	X	X	X	X	X	X	X	X	X
Patty	X	X	X	X	X	X	X	X	X	X
Glorio/a	X	X	X	X	X	D				
Popis	X	X	X	D						
Barrigo/a	X	X	X	X	X	X	X	X	X	X
Kiko	X i	X i	X i	X	X	X	X	X	X	X
Chavo/a		X (B) i	X i	X i	X	X	X	X	X	X
Jaimito/a					X (B) i	D				
Godino/az			X (B) i	D						
Sesenta-y-uno/a			X (B) i	D						
Group size sub-total:	12	13	15	12	13	11	11	11	11	11
Total average:	12									
SD(±):	1.3									

Abbreviations: X=Presence; B=Born; D=Disappeared; i=Infant (on mothers' backs).

Proportion /month:	Mar	Apr	May	Jun	Jul	Aug	Sep	Oct	Nov	Dec	Average	SD	Min	Max
Group size:	12	13	15	12	13	11	11	11	11	11	12	1.33	11	15
♂ adult	0.25	0.23	0.20	0.25	0.23	0.27	0.27	0.27	0.27	0.27	0.25	0.02	0.20	0.27
♀ adult	0.33	0.31	0.27	0.33	0.31	0.36	0.36	0.36	0.36	0.36	0.34	0.03	0.27	0.36
Juvenile/subadult	0.33	0.31	0.27	0.33	0.38	0.36	0.36	0.36	0.36	0.36	0.34	0.03	0.31	0.38
Infant	0.08	0.15	0.27	0.08	0.08	0.0	0.0	0.0	0.0	0.0	0.13	0.09	0.0	0.27

Table 3.8. Proportional composition of La Yunai group at La Suerte Biological Field Station between March-December 2006.

Field site, country, study date	Group size	Proportion ♂ adult	Proportion ♀ adult	Proportion Juvenile/subadult	Proportion Infant	Reference
Barro Colorado Island, Panama (1966–1970)	15	0.17	0.31	0.36	0.16	Oppenheimer (1982)
Barro Colorado Island, Panama (1986–1988)	20	0.19	0.32	0.37	0.12	Mitchell (1989)
Trujillo, Honduras (1980–1981)	14	0.21	0.24	0.39	0.16	Buckley (1983)
Parque Nacional Santa Rosa, Costa Rica (1972)	17	0.29	0.35	0.24	0.12	Freese (1976)
Parque Nacional Santa Rosa, Costa Rica (1983–1999)	15.8	0.21	0.31	0.34	0.13	Jack and Fedigan (2001)
Estación Biológica La Suerte, Costa Rica (2006)	12	0.25	0.34	0.34	0.13	Urbani 2009, this study (Table 3.8)
Average	15.6	0.2	0.3	0.3	0.1	
SD±	2.7	0.04	0.04	0.05	0.02	
Min.	12	0.2	0.2	0.2	0.1	
Max	20	0.3	0.4	0.4	0.2	

Table 3.9. Proportional composition of different groups of *Cebus imitator* (data from Fragaszy *et al.* 2004).

Data Collection

Research Design: Natural Field Study (Behavioral-Ecological Phase)

Quantitative behavioral data on capuchins were collected during 12 consecutive days per month (from March to August and December; and 10 days in October) over an 8-month period. The data analyzed in Chapter 4 and Chapter 5 were recorded during the 8-month study period. Data collection across consecutive days was required in order to test a series of hypotheses concerning spatial representation, foraging strategies, patterned use of feeding/resting sites, decision-making, and cognitive ecology. Behavioral data on focal animals were recorded using a 2-minute instantaneous sampling technique (Altmann 1974, Garber and Rehg 1999, Martin and Bateson 2000). Each 2-minute sample represents an individual activity record (IAR). The focal animals were adult males and adult females. The identity of the focal animal was rotated randomly each day in order to obtain a relatively equal number of observations for each of the eight adult group members. The focal animal was followed continuously from dawn to

Behavioral category	Brief definition
Feeding	Mouth processing and ingestion of any food item; including handling of a food item that is specifically intended to be directly introduced into the mouth. The food items include petiole (leaf shoot), petal (flower), fruits (and their annotated parts: seed [whole seed], Embryo/Endosperm [seed kernel; in the case of *Dipteryx panamensis* only], Endocarp [seed cover], Mesocarp [flesh], coconut apex [fleshy part around the contact area of the coconut with the palm stem]), invertebrates (Araneae: spider; Coleoptera: beetle, Hemiptera: cicada only; Homoptera: leafhopper, Hymenoptera: bee, Isoptera: ant only, Lepidoptera: butterfly, Orthoptera: grasshopper, Scorpions: scorpion), vertebrates (birds, lizards, and squirrels), and termitaria earth.
Foraging	Movement within or between single or multiple trees searching for food, both visual (scan) or manual foraging.
Resting	Any phase of behavioral inactivity.
Social interactions	The social behaviors indicated as follows: • Grooming: physical contact between individuals for social purposes or for removal of dirt and/or external parasites. • Threat: persecutory behavior generally with physical contact in which an individual or group tries to harm another. • Chasing: persecutory behavior generally without physical contact in which an individual or group tries to expel another. • Playing: an amusing behavior mostly with the presence of young individuals. • Huddling: associative behavior in which two or more individuals press against each other, making close physical contact. • Fur-rubbing: behavior in which two or more individuals apply an external substance (mainly *Citrus* spp., Rutaceae) to their bodies • Food-sharing: behavior in which an individual offers food to another. • Displacement: behavior in which an individual forces another to leave the place in use, in order to use it by itself. • Vocalization: behavior in which an individual produces a vocal sound directed to a conspecific or another animal. • Copulating: mating behavior between two individuals. • Hunting: behavior in which individuals try to catch prey, whether successful or not. The age/sex of individuals were annotated with each interaction.
Traveling	Continuous movement within multiple trees without searching for food.
Unknown	Inability to observe any behavior displayed by the focal animal.

Table 3.10. Behavioral categories recorded from the capuchins' activities at La Suerte Biological Field Station.

dusk. I was accompanied by a local field assistant to insure continuous contact with the focal animal and to aid in marking and mapping trees used for feeding and resting. The monkeys' sleeping site was located each afternoon in order to resume behavioral sampling the following day. Once the focal animal was located at the sleeping site in the morning, it was followed from ~5:45 am, until entering the sleeping site at approximately 5 pm. The length of the capuchin activity period remained constant throughout the year (mean hours=10.3±0.6) despite changes in day length from 9.4 hours in July to 11.2 hours in June.

In order to obtain detailed data on ranging behavior and activity budget, each IAR includes information on the focal animal's location (based on a grid system, see below), time of day (2-minute intervals), age/sex identity, and behavioral activity such as resting, feeding, traveling, foraging, social interaction, or unknown (Table 3.10). When the capuchins entered a tree to feed or forage, the type of food consumed was recorded. Table 3.10 includes the definitions of the behavioral categories recorded.

A **detailed grid system** of 10 × 10 m quadrats was carefully mapped and marked with surveyor's fluorescent flagging tape throughout the group's entire range. Surveyors' flagging tape also was used for marking the trail points. Each of these grid marks was assigned a unique identification number according to a coordinate system (X, Y), or a letter and a consecutive number in the case of the trail markers. The grid was made with the assistance of a professional surveyor (Franco Urbani), using a transit compass mounted in a tripod and a laser rangefinder that was tested for accuracy in this forest during the Pilot Study (see below). A total of 1,818 grid points, and 685 trail and topographic polygon points, were placed in the "small forest" of the EBLS. The grid system was used to create a detailed digital field map that included physical features of the environment such as creeks and ravines, using the ArcGIS© mapping software. All field location data were transformed into UTM coordinates in order to add them to the digital map. This map also included a detailed section of the areas used by the capuchins outside the EBLS property (see Appendix). The map serves to (a) determine the precise location of the focal animal, (b) to record travel paths used by the capuchins, and (c) determine the accurate distribution of resting, feeding, and sleeping sites. The data used in Chapter 5 were from the area covered by the grid system and did not include the observation time during which the capuchins traveled off the EBLS property.

All feeding/resting trees used by capuchins (n=306) were identified (local and scientific names, Urbani 2009: Appendix F), marked with a single identification number (using a double-faced aluminum tag and red surveyor's flagging tape), and then mapped with respect to the closest grid point (angle and distance). In the cases of 137 feeding/resting trees which were located outside the grid system, UTM measurements were recorded using close topographic polygon points or a GPS device. All feeding/resting trees visited by the capuchins were plotted on the digitized field map (total= 306 feeding/resting trees; 12 sleeping sites [10 trees and 2 bamboo tangles]). The diameter at breast height (DBH), crown diameter (by recording two radius measurements), crown height (height of the first tree branch, whenever possible using a digital clinometer/height calculator), crown shape (cone, ellipsoid, rectangle, square, spherical) and tree height (maximum height of the tree, whenever possible using a digital clinometer/height calculator) were collected for all feeding/resting sites.

Trees in which capuchins spent at least 1% of their total feeding and resting time during the field study were analyzed to calculate capuchin's "field of view" (after Garber and Jelinek 2006). Capuchin's field of view was calculated by measuring DBH ($\geq$10), crown height, and crown volume of all trees within a 20 m radius of these frequently used trees. These data were used to estimate the field-of-view or "line of sight" through the canopy in order to assess whether the capuchins could sight directly to their next feeding or resting tree.

All adult trees of two tree species used by the capuchins were identified, located, and plotted on the field map. These tree species are *Dipteryx panamensis* and *Sterculia recordiana*. These were 11 individual trees of *D. panamensis* and 6 individual trees of *S. recordiana*. These two tree species were selected because individual trees of these species were visited by the capuchins in succession. It enables me to determine whether the capuchins visited these trees using the least distance principle. In addition, these tree species were exclusively located within the EBLS, therefore they were distributed in the forested area covered by the grid system, allowing precise calculation of distances between individual trees.

Finally, weather condition (sunny; cloudy; foggy; or rainy) was recorded three times a day (6 am, 12 pm, and 4 pm) during the observation sessions. For all trees visited by capuchins that I was not able to identify directly in the field, fruits, leaves, and bark (when necessary) were collected. These samples were preserved in a standard plant press, maintained in pure alcohol, and photographed with a digital camera. These plant samples were identified by three different senior field assistants (Félix Ramón Alvarez-Malespín, Ramón Mesen-Alemán, and Elías Mesen-Alemán) and an experienced Costa Rican botanist (Reinaldo Aguilar).

Research Design: Experimental Field Study (Field Experimental Phase)

I conducted an experimental field study of capuchin spatial memory and decision-making during a 3-month period (September to November 2006). The experiments included two test conditions and were designed to examine how capuchins used nodes for navigation, and if the monkeys use a distance-minimizing principle when traveling between feeding platforms at. In addition, I evaluate which factors such as distance and resource quantity influenced capuchin foraging decisions. Experiments 1 and 2 lasted 24 consecutive days (five days of pre-baiting) and 43 consecutive days (12 days of pre-baiting) respectively and were terminated once the capuchins reached a total of 80 platform visits per experiment. Experiments consisted of four feeding platforms each measuring 2 × 2 m and raised 1.5 m above the ground. Feeding platforms were wooden structures identical in form, color, and shape. The wood selected for the platforms was odorless to human perception and no chemical products were used to protect them in order to avoid unintended olfactory cues. Platforms were located within an 8.2 ha of primary forest covered by the grid system and positioned such that the capuchins could access them using natural vegetation. The location of feeding platforms remained constant throughout each experiment but they were moved between experiments.

Previous experimental field studies indicated that ripe bananas (*Musa paradisiaca*) are a highly attractive resource for individuals of this capuchin group and that at distances ≥0.5 m capuchins are unable to use olfactory cues to discriminate between the presence or absence of a banana food reward (Garber and Paciulli 1997, Garber and Brown 2006). Depending on the protocol of the experiment, platforms contained either four or eight whole bananas (see below). The use of bananas also removes the possibility that exotic seeds might be dispersed in the forest.

The bananas placed on the platforms were not covered at the start of the pre-baiting period. However, once one platform was visited in each experiment, half of the banana rewards on each platform were covered with a large synthetic and realistic banana leaf (68 cm x 31 cm) and half were visible. This was done to stimulate manual or visual inspection of the platforms. The bananas were obtained from two plantations located within 6 km of the field site and were available throughout the year. Platforms were provisioned at dark one time per day (4:30 am), which was prior to the capuchins leaving their sleeping sites. The location of the feeding platforms, on the left bank of La Suerte River, was more than 150 m from a capuchin sleeping site (right bank of La Suerte River). The bananas were individually washed in clean water early in the morning before they were brought into the platforms, and placed in a sealed black plastic bag before being taken to the forest. All bananas remaining on a platform at the end of the day were removed once the capuchins entered their sleeping site. In addition, in the afternoon the plastic banana leaves were cleaned, and any banana skins and remaining banana pulp were collected, placed in a plastic bag, and disposed of, at the field camp.

The average weight of husked bananas used in the experiments was 118 g (complete banana weight = 180±30 g, [n=341]; banana peel weight = 63±8 g [n=147]). In order to ensure that independently of the platform location and direction of arrival the capuchins encountered identical configurations on each platform, the bananas and the plastic banana leaves were placed in the same position and orientation on each platform. None of the few banana and plantain plants that are located around the "small forest" was visited by capuchins during the 8-month behavioral-ecological phase of this research or during the experimental phase of this study.

During the field experiments, two trained field assistants equipped with short-wave radios, binoculars, and synchronized watches assisted in the research. Capuchins were followed from their sleeping sites until they arrived in the vicinity of a feeding platform. After the capuchins left the feeding platform they were followed during the rest of the day as was done during the behavioral-ecological study. The behavior of the capuchins at the feeding platform was observed from a distance of approximately 10 m, with the researchers concealed in understory vegetation. This was done in order to avoid the researchers´ providing any unintended cue as to the location of platforms. Routes capuchins travel between the feeding platforms were recorded by collecting the sequence of all grid points crossed when moving from one platform to another. Data recorded include: (a) the identity of the focal capuchin (and other capuchins) arriving at a platform, (b) the number of bananas on the platform at the time of arrival, (c) the time and direction of travel of capuchins arriving at the platform, (d) time of departure, (e) the code of platforms visited, (f) the success of the animal in obtaining a food reward, (g) amount of food consumed, (h) location where the focal animal consumed the reward (on the platform or nearby tree), (i) identity of the closest individual (≤2 m) to the focal animal, and (j) food calls and social interactions (e.g. food sharing) at the platform.

In order to evaluate the behavior of the capuchins at the platforms, two categories were evaluated. A *platform visit* was defined as the direct inspection of a platform by the focal animal whether it contained a food reward or not. This included direct-manual contact or visual inspection (visual scans). Direct-manual contact occurred when the focal animal landed on a platform or hung from a substrate that permitted the forager to obtain the food reward (≤0.5 m) (see Garber and Dolins 1996, Bicca-Marques 2000). A visual scan or visual inspection occurred when the capuchin looked directly at the platform from a distance of <10 m but did not approach the platform. During a platform visit, the identity of all capuchins on the platform was recorded. A total of 82 and 80 platform visits were recorded from Experiments 1 and 2, respectively. A *visit sequence* was defined when the focal animal was inspecting three or four platforms in succession. Data collection on visit sequences did not begin until the capuchins had visited the locations of all four experimental platforms. No other primate species visited the platforms during the study. On three occasions, one unidentified bird and two tayras (*Eira barbara*) were observed on the platforms. However, these cases occurred late in the afternoon after the capuchins had visited the platforms and left for their sleeping sites.

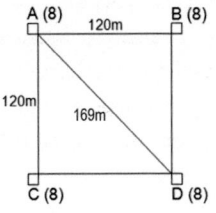

Figure 3.4. Experiment 1: Equal distance/Place constant/ Equal amount of food reward (Vertices and letters: Platforms and their codes; in parentheses: Amount of food reward).

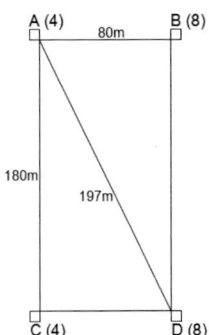

Figure 3.5. Experiment 2: Different distances/Place constant/Higher *vs.* lower food reward (Vertices and letters: Platforms and their codes; in parentheses: Amount of food reward).

Experiment 1 (Equal distance/Equal amount of food reward; square configuration, Figure 3.4). The goal of this experiment was to examine capuchin decision-making under conditions in which there was an equal amount of food on each platform, and the platforms were located in a square configuration. The research design included four platforms, each with eight bananas. The condition of this experiment remained constant for 24 consecutive days. Each platform was located 120 m distant from its neighboring platform (169 m between opposite vertices). If the distance is a primary factor in capuchin foraging decisions and the capuchins can accurately encode distance and location in an internal spatial representation, I expect the forager to visit the platform in a sequence such as A→B or A→C, but not A→D. I calculated the circuity indices of the route segments prior to visiting each platform (see below).

Experiment 2 (Different distances/Higher *vs.* lower food reward; rectangular configuration, Figure 3.5). The goal of this experiment was to identify the degree to which capuchins used both distance and quantity information in selecting feeding sites (more distant and productive feeding sites *vs.* nearer less productive feeding sites). The research design presented to the forager included 4 platforms in new locations. Two of the platforms contained four bananas each and two contained eight bananas each. The platforms were arranged in a rectangular configuration (80 × 180 m; 197 m between opposite vertices). If the quantity is the most salient cue in capuchin foraging decisions, I expect capuchins will first visit a platform that contained eight bananas and then move directly to the more distant platform containing eight bananas. If reducing distance is a more important factor in decision-making, then it is expected that capuchins will travel to the nearest feeding platform regardless of the amount of food reward, and then go from there to the nearest platform.

In both experiments, I compared the distance that the capuchins traveled to experimental feeding sites with the shortest straight-line distance between feeding platforms.

Data analysis

The behavioral data (collected as individual activity records, IAR) during the natural field study were analyzed using non-parametric statistical tests. Non-parametric tests were used because the data violate assumptions of randomness. P value is set at 0.05 and 0.01. The circuity index (CI), or distance the capuchins traveled between successive feeding sites divided by the straight-line distance between these same feeding sites, was used to test the hypotheses of route-based and coordinate-based spatial representation. When CI=1.0, a traveler/forager is taking the most direct route (Garber and Hannon 1993). Z-tests and t-tests were performed to compare means in two samples (Fowler *et al.* 1998). Chi-square (χ^2) and G_{adj} tests were used in the analysis of frequency tables of two or more categories (Sokal and Rohlf 1995), and used to determine if the pattern of platform choices selected differed from a random distribution. StatGraphic™ and Excel installed with PopTools were used as the programs to run the statistical analyses.

Travel paths were examined by plotting the 2-minute by 2-minute movement patterns of monkeys based on the location of the focal animal position. Using the ArcGIS™ map, I identified route segments and crossing points ("nodes") that were used each month. A node is defined as a circular area of 30 m in diameter (706.5 m²) that was used as a crossing point to intersect re-used routes. A route segment is defined as a path re-used by the capuchins that was bounded by two nodes. Thus, a crossing point is a fixed point in which capuchins re-orient the direction of travel (Garber 2000). Major feeding/resting trees are defined as the ones where the monkeys spent a minimum of 1% of their total feeding and resting time (after Jelinek and Garber 2006). To plot the directions from which these trees were visited I used GEOrient© v9.x. Using AutoCAD™ and CorelDRAW™, I calculated the total distance of the routes. ArcGIS™ was used to (a) plot the distribution of feeding and resting trees visited by capuchins, (b) identify the location and types of vegetation zone in the capuchins´ home range, (c) plot the location of the grid points placed in the "small forest", (d) plot the location of feeding platforms used in Experiment 1 and Experiment 2, (e) locate physical features of the environment such as creeks, and river and forest borders, and (d) plot the position of forest trails. These maps are found in Figures 6.2 and 6.4 as well as Appendices.

Pilot study

During the summer of 2004, I conducted a 1-month pre-dissertation pilot study at EBLS (Urbani 2004). At that time the La Yunai group was followed for two consecutive weeks, and preliminary data on ranging, feeding, and social behavior were recorded using a 2-minute instantaneous focal animal sampling technique. In addition, the locations where experimental platforms would be placed were evaluated. Moreover, during this period, methodologies such as techniques of observations and monkey identification in this forest, familiarization with the characteristics of the forest, and the use of topographic materials and other field equipment were tested.

Ethical statement

The project complied with non-human primate protection rules established by the University of Illinois' Institutional Animal Care and Use Committee (IACUC-UIUC). The IACUC protocols for this project were approved on May 10th, 2005 (Protocol number: 05070). No handling and/or trapping of capuchins were done during this field research.

B. Urbani declares no conflict of interest. This research follows Costa Rican regulations as well as the codes of ethics of the American Society of Primatologists and the International Primatological Society.

Data availability

The data used in this research is found in Urbani (2009: 249–355) (see Table 3.11). Urbani (2009) can be downloaded on the webpage of ProQuest (www.proquest.com). In addition, all data used for analyses are available upon request.

Appendix in Urbani (2009)	Heading of the appendix
Appendix D	Daily routes of the La Yunai group during the natural field study in 2006.
Appendix E	Monthly routes of the La Yunai group during the natural field study.
Appendix F	Monthly nodes and route segments used by the La Yunai group during the natural field study.
Appendix G	Detailed pattern of tree visits used by the La Yunai group during the natural field study.
Appendix H	Individual trees used by the La Yunai group.
Appendix I	Routes used by the La Yunai group during the experimental field study.
Appendix J	List of equipment used in the field.
Appendix K	Map of the complete area used by the La Yunai group.

Table 3.11. Open data availability of this study (appendices in Urbani 2009).

4

The Behavioral Ecology of a Group of Wild White-Faced Capuchin Monkeys (*Cebus imitator*)

This chapter examines the behavior and ecology of a group of white-faced capuchin monkeys (*Cebus imitator*) inhabiting a tropical rainforest in Costa Rica. Specifically, I studied the capuchins´ activity budget, diet, and ranging patterns. In addition, I compared these data with published datasets of other populations of white-faced capuchins, and other capuchin species. I also empirically explore ideas on resource seasonality, foraging rules, feeding ecology, and the use of keystone and fallback foods in this capuchin group.

Resource seasonality is defined as the fluctuation in the availability of food items during an annual cycle (Clutton-Brock and Harvey 1977, Peres 1994, van Schaik and Brockman 2005). Resource seasonality impacts diet (Terborgh 1983), nutrition, activity budget (White 1998, Knott 2005), within and between group social interactions (van Schaik *et al.* 2005), and foraging and reproductive strategies (Janson and Verdolin 2005, Brockman and van Schaik 2005). For example, Knott´s (1998, 2005) long-term study on diet and seasonality in Bornean orangutans found that during the month of highest fruit availability (January), these apes ingested fruits that were more than five times richer in calories and higher in carbohydrates and lipids than those consumed during periods of relative food scarcity (May). In terms of reproductive ecology, Janson and Verdolin (2005) compared birth peaks in relation to seasonality in 60 species of prosimians and monkeys. Seasonality in this study was evaluated using a global climatic dataset that included temperature and rainfall, and published information on phenological data of 28 sites with wild primate populations. These authors found that the strongest predictor of birth peaks and infant survivorship was fruit availability and abundance. These two variables explained nearly two-thirds of the variance in the dataset. These results must be viewed with caution, however, due to the fact that different authors used different definitions of resource seasonality, alternative measures of fruit availability, and different criteria for infant survivorship. However, Janson and Verdolin (2005) argue that across these species increased birth rates are generally correlated with times of the year during which the production of fruit resources was expected to be high.

Food distribution and availability have been argued to be fundamental factors affecting the feeding and foraging habits of wild primates (Garber 1987, White 1998, Ganzhorn *et al.* 2003). In seasonal environments or during periods of resource scarcity, primate species such as capuchins (*Sapajus apella* and *Cebus albifrons*), hamadryas baboons (*Papio hamadryas*), black-and-white colobus (*Colobus angolensis*), gibbons (*Hylobates agilis, H. mulleri*), sifakas (*Propithecus diadema*), lemurs (*Eulemur fulvus* and *E. rubriventer*), and tarsiers (*Tarsius spectrum*) are reported to exploit a smaller number of plant species (Terborgh 1983, Yamashita 1996, Lowe and Sturrck 1998, Gursky 2000, McConkey *et al.* 2003, Swedell *et al* 2008). In some cases, this is associated with the increase in day range. In other cases, this is associated with a decrease in day range and more intense exploitation of a core area. For example, in hamadryas baboons (*Papio hamadryas*) during the dry period of the year (October to May) day range increased from 8,500 to up to 13,000 m (Kummer 1968). Increased day range was associated with the scattered distribution of a limited number of available feeding sites and water sources (Kummer 1968). However, during the dry season, other primate species such as mustached tamarins (*Saguinus mystax*) and saddleback tamarins (*Leontocebus fuscicollis*) in Amazonian Peru, were able to exploit foods such as nectar and exudates with only minor changes in travel distance (*S. mystax*: day range = 2166 m in dry season, day range = 2013 m in the wet season; *L. fuscicollis*: day range = 2125 m in the dry season, day range = 1983 m in the wet season) (Garber 1993). The author suggested that tamarins accomplished this by maintaining a year-round foraging pattern in which several trees of a small number of target tree species were exploited intensively over the course of a given one-month or two-month period. As new trees began fruiting, flowering, or producing exudate, the tamarins incorporated them into their foraging pattern. Other species such as squirrel monkeys (*Saimiri sciureus*) also are reported to exhibit a relatively consistent daily path length during dry and wet seasons (Stone 2004). This is accomplished by responding to periods of fruit scarcity with a change in dietary emphasis by focusing more on insect prey.

Two proposed concepts central to dietary seasonality are keystone resources and fallback foods. **Keystone plant resources** (hereafter referred to as KPRs) are described as a limited number of plant species that provide a large amount of food to a broad number of animal taxa inhabiting a forest community, especially during otherwise food-limited times of the year (Terborgh, 1986, Peres 2000). Terborgh (1986) suggests that figs, palm nuts, and nectar represent KPRs for many primate species. For example, at Cocha Cashu in the Peruvian Amazon, mammals such as marsupials, primates, and procyonids, as well as birds like toucans, trumpeters, guans, and passerines consumed fig species during several months of the year. Figs fruit asynchronously, produce a large crop and therefore are be available to consumers during times of the year when other fruits are not available.

Palm nuts are consumed by peccaries and capuchin monkeys during the dry season months of May and June in southwestern Peru (Terborgh 1983). Palm nuts represent a rich source of lipids. Terborgh (1983) reported that the density of palms (*Scheelea* and *Astrocaryum*) at Cosha Cashu was extremely high, averaging 25 and 39 individuals per hectare. Brown capuchins (*Sapajus apella*) are reported to have morphological adaptations of their teeth and jaws that permit them to exploit and access these hard nuts (Wright 2004). In contrast, white-fronted capuchins (*C. albifrons*) consume palm nuts during food-limited times of the year by selecting those that were infected with beetle larvae, which made them softer and thus, easier to break. Terborgh (1986) proposed a model in which KPRs are distributed in relatively low density, but provide resources critical for maintaining the vertebrate community. The limitations of this concept as defined by Terborgh (1986) include the assumptions that only animals with particular anatomical adaptations can exploit KPRs, and that only a very limited subset of plant genera may fulfill the role of KPRs in tropical forests (Terborgh 1986).

Peres (2000) challenged the concept of KPRs based on his observations of the feeding behavior of parrots, primates, peccaries, and ungulates on exudates from the pods of *Parkia* trees. He suggests that KPRs need to be evaluated and defined in relation to the presence of other resources available to vertebrate consumers. Secondly, he argued that consumer specificity is a critical feature of KPRs. Peres (2000) defines consumer specificity as the set of consumers that rely consistently on KPRs from year to year. Peres (2000) proposed that KPRs vary in terms of their year-to-year reliability in the timing and production of fruits, flowers, and exudates. He concluded that the ecological concept of a KPR is more appropriate when used for a given forest in a given year rather than an evolutionary concept in which a large set of vertebrate species has specific adaptations to exploit a small set of plant species. Finally, as pointed out by Peres (2000) the concept of KPRs tends to be narrow by limiting the number of plant species that may function as critical food resources for several species of vertebrates.

Building on the concept of KPRs, Marshall and Wrangham (2007) have suggested that during periods of food scarcity, primates may rely on a set of resources termed **fallback foods** (hereafter referred to as FBFs). The authors defined FBFs as low-quality resources that are consumed when higher-quality foods are unavailable. FBFs are expected to account for only a small percentage of yearly feeding time. The authors suggest that FBFs are "[food] items assumed to be of relatively poor nutritional quality and high abundance, eaten particularly during periods when preferred foods are scarce… Accordingly, one can operationally define FBFs as foods whose use in negatively correlated with the availability of preferred foods" (Marshall and Wrangham 2007: 1220). Marshall and Wrangham (2007) offered a critical function argument by suggesting that in order to process seasonally available low-quality foods, primates have evolved specific digestive, dental, and/or masticatory adaptations.

Lambert (2007) argued that grey-cheeked mangabeys (*Lophocebus albigena*) evolved thick molar enamel to exploit hard foods such as bark and seeds. During El Niño years in which the most commonly consumed foods were not available, mangabeys relied on hard-to-process resources with high crushing resistance (1.81 kg/mm²). Lambert (2007) suggested that the thick molar enamel of mangabeys evolved in response to the challenges posed by hard food items that served as FBFs during certain years.

Marshall and Wrangham (2007) distinguished two types of FBFs. Filler FBFs comprise a small part of an individual's diet during a food-limited period of the year. According to Marshall and Wrangham (2007 citing Terborgh [1983], Davies *et al.* [1998], respectively), brown capuchins (*Sapajus apella*) relying on palm nuts in southwestern Peru, and leaf monkeys (*Presbytis melalophos*) selecting mature leaves in the Sandakan Peninsula of Malaysian Borneo represent examples of filler FBFs. However, this is not the case, as Terborgh (1986) provided evidence that palm nuts represent a major food source for capuchins between April and August and accounted for ~60% of feeding time. The same is true for *Presbytis melalophos* for which mature leaves accounted for 8.0% of the diet year-round. These examples highlight weaknesses in Marshall and Wrangham´s (2007) framework, namely nuts represent common foods for capuchins during half of the year, and mature leaves were evenly used throughout the year by langurs.

Staple FBFs are defined by Marshall and Wrangham (2007) as foods that are available and eaten throughout the year, but are consumed most intensively when other resources are scarce. Staple FBFs may account for up to 100% of the diet during periods of food scarcity, and over short periods of time provide the majority of the energetic requirements of the consumer. Marshall and Wrangham (2007 citing Chapman [1987], Pwzyk and Mowry [2003]) suggest that the ingestion of young leaves by mantled howler monkeys (*Alouatta palliata*) and indris (*Indri indri*) are examples of staple FBFs. However, in mantled howlers, Milton (1980) reported that young leaves are eaten during both dry and wet seasons, accounting for 48.9% of the feeding time in the wet season and 35.2% in the dry season. Thus, young leaves are consumed throughout the year and represented commonly ingested food items. Similarly, Kowalewski (2004) found that in northern Argentina gold-and-black howler monkeys (*Alouatta caraya*) devoted ~40% of their feeding time to young leaves. This site is located at the southern extreme of the distribution of Neotropical primates, and is subject to extreme variation in temperature, rainfall, and resource seasonality. The pattern of consuming young leaves year-round by *Alouatta caraya* does not fit the model of staple FBFs.

In a recent paper, Porter *et al.* (2009) critique Marshall and Wrangham's (2007) use of FBFs. Marshall and Wrangham (2007) argued that the properties of FBFs rather than commonly consumed foods have selected for anatomical and behavioral adaptations of the consumer. Porter *et al.* (2009) found that exudates exploited by callimicos (*Callimico goeldi*) during the dry season in northwestern Bolivia were characterized by high levels of structural carbohydrates. They argued that the ability of callimico to digest exudates, a low-quality resource during the dry season, was facilitated by their ability to digest fungi, a low-quality resource that is consumed throughout the year. Porter *et al.* (2009) stated that the digestive adaptations that permit the consumption of fungi during the annual cycle also may facilitate the use of other difficult-to-digest resources such as exudates during periods of fruit scarcity.

An alternative approach to the study of primate diets focuses on commonly exploited foods. For example, Sussman (1987) has argued that, although primates may consume foods from >50 plant species during the year, many primate species rely mainly on a small subset of plant species for the majority of their diet. I refer to these foods as **highly selected foods** (hereafter referred to as HSFs). HSFs can be defined as those food resources that comprise ≤10 plant species in the forest but account for ≥50% of the total monthly and yearly plant feeding time. HSFs generally are available year-round. Efficient exploitation of HSFs requires cognitive skills associated with spatial memory to locate and relocate a small set of critical feeding sites.

In this chapter, I examine concepts of KPRs, FBFs, and HSFs as they relate to the feeding behavior of a group of wild white-faced capuchin monkeys in northeastern Costa Rica. Questions addressed are: (a) what is the activity budget, diet, and day range of capuchins at EBLS?, (b) does the activity budget, diet, and day range vary significantly across different months of the year (stability *vs.* variability)?, (c) does white-faced capuchin behavior and ecology at EBLS differ from the behavior and ecology of white-faced capuchins living in sites characterized by a pronounced dry season?, (d) what set of ecological factors account for inter-site variability in white-faced capuchin behavior, ecology, and diet?, (e) is white-faced capuchin feeding behavior best understood in terms of the use of KPRs, FBFs, or HSFs?

	Total recorded bouts	Mean capuchin activity hours per day	Observation hours during the month	Mean observation hours per day	Complete observation days out of total follow days
March	3724	10.3	124.1	10.3	12 out of 12
April	3401	9.5	113.4	9.5	12 out of 12
May	3774	10.5	125.8	10.5	12 out of 12
June	4047	11.2	134.9	11.2	12 out of 12
July	3126	9.4	104.2	8.7	10 out of 12
August	3933	10.9	131.1	10.9	12 out of 12
September (Exp. 1)	7224	10.3	240.8	10.0	23 out of 24
October	3157	10.5	105.2	10.5	10 out of 10
October (Exp. 2)	6417	9.8	213.9	9.7	20 out of 22
November (Exp. 2)	5544	9.8	184.8	8.8	17 out of 21
December	2408	10.3	80.3	6.7	6 out of 12
Total (Beh-Ecol. days)	*27570*	*10.3*	*98.0*	*9.8*	*86 out of 94*
SD (±)	-	*0.6*	-	*1.5*	-
Min.	-	*9.4*	-	*6.7*	-
Max.	-	*11.2*	-	*11.2*	-
Total (Exp. days)	19185	10.0	213.1	9.5	60 out of 67
Total	46755	10.2	155.6	9.7	146 out of 161
Total obs. hours (Beh-Ecol. days)	*919.0*				
Total obs. hours (Exp. days)	639.5				
Total observation hours	1558.5				

Note: Averages from the behavioral-ecological study months are indicated in italics.

Table 4.1. Observation information from the La Yunai group during 2006.

Materials and Methods

This research was carried out at Estación Biológica La Suerte (EBLS), northeastern Costa Rica (10°26'N; 83°47'W). The data presented in this chapter come from 12-consecutive day follows per month between March-August 2006 and December 2006, as well as a 10-consecutive day period in October 2006, totaling 8 observation months (the total averages from these behavioral-ecological study months are indicated in italics within the tables of this chapter). The data consisted of 919 observation hours (27,570 activity records taken every 2 minutes [Table 4.1]). During an additional 3-month period (September-November 2006), I also conducted the experimental field phase of this research. For that period I collected information on the diet, feeding/resting trees visited, and sleeping sites used by the capuchins (639.5 hours, 19,185 activity records). Table 4.1 summarizes this information. A detailed description of the Materials and Methods is presented in Chapter 3.

	Feeding	Foraging	Resting	Social interactions	Traveling	Unknown
March (n=3724)	32.6 [1]	41.5	4.5 [2,3]	3.7	16.2	1.4
April (n=3401)	24.1	47.8 [4]	9.1 [5,6]	5.0	12.5	1.6
May (n=3774)	21.4	46.1 [7]	13.2 [8,9]	2.5	15.4	1.5
June (n=4047)	23.9	50.3 [10]	9.0	1.4	14.7	0.69
July (n=3126)	14.0	60.5	6.4	1.3	16.1	1.7
August (n=3933)	22.7	49.3 [11]	5.5 [12]	3.5	18.6	0.43
September (Exp. 1) (n=7224)	15.1	56.7 [13]	10.6 [14]	2.2	14.7	0.82
October (n=3157)	10.3	62.7 [15]	10.6 [16]	4.2	11.6	0.67
October (Exp. 2) (n=6417)	13.1	56.6 [17]	10.6 [18]	4.6	14.6	0.51
November (Exp. 2) (n=5544)	12.6	61.5	4.9	3.3	16.8	0.90
December (n=2408)	13.7 [19]	62.5 [20]	4.2 [21]	3.1	16.0	0.58
Total (Beh-Ecol. days) (n=27281)	*20.3 (n=5603)*	*52.6 (n=14208)*	*7.8 (n=2159)*	*3.1 (n=849)*	*15.1 (n=4169)*	*1.1 (n=293)*
SD (±)	*7.3*	*8.2*	*3.4*	*1.3*	*2.2*	*0.51*
Min.	*10.3*	*41.5*	*4.5*	*1.3*	*11.6*	*0.43*
Max.	*32.6*	*62.7*	*13.2*	*5.0*	*18.6*	*1.7*
Total (Exp. days)	13.6	58.3	8.7	3.4	15.4	0.74
Total	17.0	55.4	8.3	3.2	15.3	0.90

Note: The data are presented in percentages. Averages from the behavioral-ecological study months are indicated in italics. *Comments:* [1] One recorded bout for nursing an infant while the focal adult female was feeding; [2] One recorded bout of self-rubbing of a liana leaf (*Philodendrum* spp., Aracea); [3] Three recorded bouts of self-grooming; [4] One recorded bout for nursing an infant while the focal adult female was foraging (one juvenile was observed *ad libitum* using the lemon odor-like fruit of *Siparuna* spp., Monineacea); [5] 10 recorded bouts of self-grooming; [6] One recorded bout of self-rubbing with a liana leaf (*Philodendrum* spp., Aracea); [7] One recorded bout for drinking water from the river; [8] Two recorded bouts of self-grooming; [9] One recorded bout of self-rubbing of a liana flower (*Singonium schottianum*, Araceae); [10] One recorded bout of drinking water from a hole in a tree; [11] One recorded bout for nursing a juvenile while the focal adult female was foraging; [12] Four recorded bouts of self-grooming; [13] One recorded bout for nursing a juvenile while the focal adult female was foraging; [14] One recorded bout of self-grooming; [15] One recorded bout for nursing a juvenile while the focal adult female was foraging ; [16] Four recorded bouts of self-grooming; [17] One recorded bout of drinking water on the river; [18] Ten recorded bouts of self-grooming; [19] Eight recorded bouts of drinking coconut water; [20] One recorded bout of self-grooming; [21] Two recorded bouts of urine washing

Table 4.2. Activity budget (%) of the La Yunai group during 2006.

Results

Activity budget

The group was active an average of 10.3±0.6 hrs/day (min: 9.4 hrs/day in July and max: 11.2 hrs/day in June; Table 4.1). There were no significant differences across months in the total number of hours/day the capuchins were active (χ^2=0.29, d.f.=7, p>0.05). White-faced capuchin monkeys at EBLS spent the majority of the time foraging (Table 4.2). As indicated in Table 4.2, foraging accounted for over half of the capuchin daily activity budget (52.6±8.2%). Feeding accounted 20.3±7.2%, and traveling 15.1±2.2%. The amount of time that capuchins engaged in each behavioral activity remained consistent across months. Controlling for monthly differences in the number of hours the group was active per day, only feeding varied significantly across months (Table 4.2, 4.3). During the month of March, feeding accounted for 32.5% of the total activity budget. This was associated with an increase in time spent consuming *Dipteryx panamensis*. For example, in May and

Behavioral activity	d.f..	χ2	P
Feeding	7	18.18	<0.05*
Foraging	7	8.86	>0.05**
Resting	7	9.04	>0.05**
Social interactions	7	3.86	>0.05**
Traveling	7	2.26	>0.05**
Unknown	7	1.72	>0.05**

Table 4.3. Statistical comparison of time invested in behavioral activities during the behavioral-ecological study months.

Note: * Significantly different; ** Not significantly different.

	Gro	Voc	Play	Thr	F-Sha	Cha	Hud	Disp	Fur-Rub1	Cop	Hunt2
March (n=137)	78.1	6.6	6.6	2.9	1.5	1.5	1.5	0.73	-	-	0.73
April (n=171)	38.0	12.3	8.2	1.8	-	2.3	14.0	-	5.9	12.9	4.7
May (n=96)	60.4	24.0	2.1	4.2	1.0	1.0	2.1	-	5.2	-	-
June (n=56)	55.4	8.9	3.6	14.3	3.6	1.8	-	.	8.9	3.6	-
July (n=40)	70.0	10.0	-	-	-	15.0	-	-	-	-	5.0
August (n=137)	49.6	14.6	2.9	-	0.73	2.2	23.4	-	2.9	-	3.7
September (Exp. 1) (n=156)	44.9	10.3	18.0	3.2	-	-	3.2	-	3.2	1.3	5.8
October (n=131)	76.3	10.7	-	3.1	-	-	4.6	-	3.1	-	2.3
October (Exp. 2) (n=6417)	67.1	3.4	9.1	1.0	-	14.4	.	-	3.4	-	1.7
November (Exp. 2)(n=184)	81.3	1.6	-	-	-	9.2	-	-	6.5	1.1	-
December (n=74)	60.8	13.5	-	-	-	-	5.4	-	-	13.5	5.4
Total (Beh-Ecol. days) (n=849)	61.1	12.6	2.9	3.3	0.85	3.0	6.4	0.09	3.3	3.7	2.7
SD (±)	13.6	5.3	2.6	5.1	1.3	5.4	8.6	-	2.5	5.6	1.8
Min.	49.6	6.6	0	0	0	0	0	0	0	0	0
Max.	78.1	24.0	8.2	14.3	3.6	15.0	23.4	0.73	8.9	13.5	5.4
Total (Exp. days)	64.4	5.1	9.0	1.4	0	7.9	1.1	-	4.4	0.79	2.5
Total	62.8	8.8	6.0	0.70	0.43	5.4	3.7	0.05	3.8	2.3	2.6

Abbreviations: Gro=Grooming; Voc=Vocalization; Play=Playing; Thr=Threat; F-Sha=Food-sharing; Cha=Chasing; Hud=Huddling; Disp=Displacement; Fur-Rub=Fur rubbing; Cop=Copulating; Hunt=Hunting
Note: The data are presented in percentages. Averages from the behavioral-ecological study months are indicated in italics. Social interactions accounted for 3.1±1.3% of capuchin[s activity budget.
Comments: [1] All fur-rubbing events were performed with the use of lemon (*Citrus aurantifolia*, Rutaceae); except in August when they were performed using bitter orange (*Citrus aurantium*, Rutaceae), in these cases, the oranges were smashed against tree branches, and the liquid used for rubbing was obtained from such branches; [2]All cooperative hunting events were unsuccessful; except the case of December on variegated squirrels (*Sciurus variegatoides*).

Table 4.4. Social interactions (%) of the La Yunai group.

October, *Dipteryx* accounted for only 14.7% and 12.8% of total plant feeding time, whereas in March and April, *Dipteryx* accounted for 46.3% and 38.9% of total plant feeding time (see next section: Table 4.11).

Table 4.4 summarizes the social interactions of the members of the capuchin group. Social interactions accounted for only 3.1% of the capuchin activity budget. Grooming was the most common social behavior (61.1±13.6%, n=519) followed by vocalizing (12.5±5.3%, n=107). A comparison of the frequency of affiliative behaviors (grooming, vocalization, playing, food-sharing, huddling, fur-rubbing, copulation, hunting; n=794) vs. agonistic behaviors (threat, chasing, displacement; n=54) indicates a significant difference (G_{adj}=1623, d.f.=9, p<0.01). Overall, affiliative behaviors accounted for 93.7% of all within-group social interactions. The rate of aggression per individual/hour was 0.0002, or approximately once per 8 months.

Diet, feeding, and foraging behavior

The capuchins devoted 57±3.8% of feeding and foraging time to plant parts (see details below), and 42.9±3.8% to animal matter (Table 4.5, not significantly different throughout the study period: χ^2=377.88, d.f.=7, p>0.05). In 62.4% of cases (234/375), I could identify the type of invertebrates exploited (Table 4.6). The capuchins devoted the majority of prey feeding to the exploitation of Orthoptera (20.6±7.5%), Hymenoptera (15±12.3%), and Isoptera (11.4±7.1%) (47%: Table 4.6) (for a recent list based on DNA analysis of the invertebrates in the diet of the capuchins of the EBLS "Large Forest", see Mallott 2016, Mallott *et al.* 2017).

I also examined patterns of insect feeding. During 93.5±18.3% of the observation days (n=94 days; Table 4.7), the capuchins fed on fruit early in the morning, and then after 8 am (66.6±18.3%, n=94 days) began foraging for insects. A pattern of early morning fruit feeding and later morning insect feeding may reflect two factors. Capuchins may require ready-to-use energy from fruits after a nighttime resting period. Alternatively, Dawson (1976, after Janzen 1973) suggested that this pattern of insect foraging, which he described in the Panamanian tamarin (*Saguinus geoffroyi*), may reflect the fact that larger-bodied insects such as orthopterans become more active as the morning temperature increases. I observed one case of geophagy on the soil of a termitarium (in April: 0.06% of feeding time).

Between March and December, the capuchins visited a total of 306 feeding and resting trees. Seventy-five percent were feeding trees (n=231, including 4 lianas on trees: *Passiflora ambigua*, Passifloraceae; 1 woody liana on a tree: *Maripa nicaraguensis*, Convolvulaceae; and 1 woody epiphyte on a tree: *Clusia* spp., Clusiaceae), and 20.3% were resting trees (n=62). Thirteen trees were used as both feeding and resting sites (4.2%). During this study, the capuchins fed on 69 plant species of 45 plant genera. *Dipteryx panamensis* accounted for 14% of total plant feeding time. Four other species also were frequently exploited by the capuchins (*Ficus americana* 11.9%, *Bactris gasipaes* 7.4%, *Ficus insipida* 6.8%, and *Inga spectabilis* 6.6%, *Nephelium lappaceum* 6.0%), accounting for a total of 38.7% of plant feeding time (Table 4.8). Overall, 65.3% of capuchin feeding time involved four tree families Moraceae, Fabaceae/Mimosaceae, Fabaceae/Papilonaceae, and Arecaceae (Table 4.9).

During each month of the study, the most commonly exploited tree species were *Dipteryx panamensis* (March, April, May, October), *Bactris gasipaes* (August, October, November), *Ficus americana/F. insipida* (April, May, June, August, September), *Inga spectabilis* (March, April, July), and *Nephelium lappaceum* (October, November). *B. gasipaes* is an introduced tree species cultivated at EBLS in the late 1980s. The capuchins visited 22 *B. gasipaes* trees, including 7 individual trees that were each visited during two consecutive months. Consuming fruits of *B. gasipaes* accounted for approximately one-third of capuchin feeding time (August: 31.8%, and October: 40.9%) (Table 4.10, Table 4.11). *B. gasipaes* trees were the most frequently visited trees in November (25.2% of plant feeding time).

D. panamensis ranked as the most commonly fed plant species during the entire study period (seven months). Fruits of *D. panamensis* accounted for 46.3% of feeding time in March (8 out of 9 individual trees visited) and 38.9% in April (4 out of 9 individual trees visited). During the month of August, only 0.4% (one out of 9 individual trees visited) of capuchin feeding time was devoted to *D. panamensis*.

Fruits of *Ficus americana* were ingested during three consecutive months, and accounted for 16.4% of plant feeding time in April, 17.5% in May, and 38.1% in June. There were 15 *F. americana* trees used in the group's home range, and three were extensively exploited each month. A second fig species, *Ficus insipida* was a major food source consumed by the capuchins in September and August (*Ficus insipida*, 60.4% of the total feeding time of plant parts). During August, the capuchins visited five of six *F. insipida* trees. In March, April, and July, *Inga spectabilis* was an important food source for the capuchins. This species exhibited an extended asynchronous fruiting pattern with the capuchins exploiting two of six *I. spectabilis* trees in March, four of six trees in April and two of six trees in July.

In sum, the top six tree species (*Dipteryx panamensis, Bactris gasipaes, Ficus americana, F. insipida, Inga spectabilis*, and *Nephelium lappaceum*) comprised 52.7% of the total plant feeding time of white-faced capuchins at EBLS (Table 4.8). However, these trees were not common within the group's home range. Based on my vegetation survey, trees of these six species occurred at a density of ≤5 individuals/ha, and *D. panamensis* at a density of 0.5 individuals/ha. In December, *Alchornea costaricensis*, which ranked ninth as the most commonly used tree year-round (Table 4.10), accounted for approximately 50% of white-faced capuchins plant feeding time.

When exploiting fruit tissues, capuchins consumed mesocarp/seed 44.4% of the time, endocarp 29.0% of the time, and the seed embryo (= *D. panamensis* kernel)/ endosperm 14.0% of the time (Table 4.12, 4.14). Endocarp was present in the diet of the capuchins during all 10 months of the study period, ranging from 52.4% of plant feeding time in October to 19.4% in March. The endocarp of *Inga spectabilis, Nephelium lappaceum*, and *Brosimum alicastrum* accounted for 16.0% of total plant feeding time. Other fruit parts such as the apex of coconuts (from a nearby plantation of *Cocos nucifera*), endocarp/ coconut apex (from *B. gasipaes*), mesocarp, and endocarp/seed were consumed less frequently and accounted for 11.4% of plant feeding time. Petal and petiole consumption accounted for 1.1% of the plant feeding time (Table 4.12). Capuchins were observed to consume the petiole of *Artocarpus altilis* only. Thus, overall, the diet of these capuchins principally involved the exploitation of insects, fruit mesocarp, and seeds. Moreover, white-faced capuchins acted both as potential seed dispensers and seed predators for the plants they consumed.

Table 4.5 (following page above). Types of items selected by members of the La Yunai group during feeding/foraging activities.

Table 4.6 (following page below). Animal matter (%) exploited by the La Yunai group during 2006.

	Feeding				Foraging			Feeding / Foraging			
	Pla.	Anim.	Term.	Total bouts	Pla.	Anim.	Total bouts	Pla.	Anim.	Term.	Total bouts
March	97.4	2.6	-	1210	30.0	70.1	1546	59.7	40.4	-	2756
April	91.6	8.3	0.06	819	43.5	56.5	1624	59.7	39.9	0.04	2443
May	93.4	6.7	-	808	42.9	57.1	1740	59.0	41.1	-	2548
June	95.9	4.1	-	967	40.0	60.0	2037	58.0	42.0	-	3004
July	91.1	8.9	-	438	44.0	56.0	1892	52.9	47.1	-	2330
August	93.4	6.6	-	892	47.2	52.8	1939	61.8	38.2	-	2831
Septemb.(Exp. 1)	91.5	8.5	-	1094	41.9	58.1	4097	52.3	47.7	-	5191
October	84.3	15.7	-	325	48.4	51.6	1978	53.5	46.6	-	2303
October (Exp. 2)	92.5	7.5	-	838	47.5	52.5	3630	55.9	44.1	-	4468
Novemb. (Exp. 2)	94.3	5.7	-	700	48.9	51.1	3409	56.6	43.4	-	4109
Decemb.	76.9	23.1	-	329	46.2	53.8	1504	51.7	48.3	-	1833
Total (Beh-Ecol. days)	*90.5*	*9.5*	*0.06*	*14208*	*42.8*	*57.2*	*14260*	*57.0*	*42.9*	*0.005*	*20048*
SD (±)	*6.16*	*6.2*	*-*	*-*	*5.8*	*3.7*	*-*	*3.8*	*3.8*	*-*	*-*
Min.	*76.9*	*2.6*	*-*	*-*	*30.0*	*52.8*	*-*	*51.7*	*38.2*	*-*	*-*
Max.	*97.4*	*23.1*	*0.06*	*-*	*47.2*	*70.1*	*-*	*61.8*	*48.3*	*0.005*	*-*
Total (Exp. days)	92.8	7.2	-	2632	46.1	53.9	11136	55.0	45.0	-	13768
SD (±)	*1.4*	*8.7*	*-*	*-*	*3.7*	*3.7*	*-*	*2.3*	*2.3*	*-*	*-*
Total	91.6	8.4	0.06	16840	43.7	56.3	25396	56.0	44.0	0.003	33816
SD (±)	*3.4*	*1.8*	*-*	*-*	*5.4*	*5.4*	*-*	*1.0*	*1.1*	*-*	

Abbreviations: Pla.: Plant parts (%). Anim.: Animal matter (%). Term.: Termitarium earth (%).
Note: Averages from the behavioral-ecological study months are indicated in italics.

	Ara	Col	He	Ho	Hy	Hy (eggs)	Iso	Lep	Ort	Sco	Unid. Inv.	Vert
March (n=29)	-	-	-	-	20.7	-	24.1	-	20.7	-	34.5	-
April (n=68)	-	1.5	-	-	25.0	-	7.4	2.9	17.7	-	42.7	2.9 [1]
May (n=53)	-	1.9	-	7.6	11.3	5.7	3.8	3.8	28.3	-	37.7	-
June (n=40)	-	-	4.9	2.4	4.9	7.3	7.3	-	33.2	2.5	25.0	10.0 [2]
July (n=39)	-	-	2.6	5.1	38.5	-	15.4	-	23.1	-	25.6	-
August (n=59)	1.7	1.7	-	3.4	6.8	-	10.2	3.4	15.4	-	55.9	1.7 [3]
September (Exp. 1) (n=93)	-	-	-	6.5	9.7	8.6	16.1	2.2	23.7	-	32.3	1.1 [4]
October (n=51)	-	2.0	-	5.9	11.8	-	17.7	7.8	17.7	-	37.3	-
October (Exp. 2) (n=63)	-	-	-	1.6	4.8	-	12.7	7.9	34.9	-	38.1	-
November (Exp. 2) (n=40)	-	-	-	10.0	10.0	-	17.5	5.0	17.5	-	40.0	-
December (n=76)	-	2.6	-	1.3	1.3	-	5.3	2.6	9.2	-	14.5	63.2 [5]
Total (Beh-Ecol. days) (n=415)	*0.2*	*1.4*	*0.9*	*3.2*	*15.0*	*1.6*	*11.4*	*2.6*	*20.6*	*0.3*	*34.2*	*9.7*
SD (±)	*-*	*1.0*	*1.8*	*2.8*	*12.3*	*3.0*	*7.1*	*2.7*	*7.5*	*-*	*12.6*	*21.9*
Min.	*-*	*-*	*-*	*-*	*1.3*	*-*	*3.8*	*-*	*9.2*	*-*	*14.5*	*2.9*
Max.	*1.7*	*2.6*	*4.9*	*7.6*	*38.5*	*8.6*	*24.1*	*7.9*	*33.2*	*2.5*	*55.9*	*63.2*
Total (Exp. days)	-	-	-	6.0	8.2	2.9	15.4	5.1	25.4	-	36.8	0.4
Total	0.1	0.6	0.3	4.6	11.6	2.2	13.4	3.8	23.0	0.2	35.4	5.0

Abbreviations: Unid. Inv. = Unidentified invertebrate; Ara = Araneae; Col = Coleoptera; He = Hemiptera; Ho = Homoptera; Hy = Hymenoptera; Iso = Isoptera; Lep = Lepidoptera; Ort = Orthoptera; Sco = Scorpion; Ver = Vertebrate. Vertebrates: [1] Unidentified young parrot; [2] Chesnut-colored woodpecker (*Celeus castaneus*) [this bird was not hunted by the focal animal, thus there were no registered "cooperative hunting" bouts in Table 4.4]; [3] Unidentified lizard; [4] Unidentified bird eggs, [5] Variegated squirrel (*Sciurus variegatoides*).
Note: Averages from the behavioral-ecological study months are indicated in italics.

	Time allocation of first invertebrate feeding bouts (after feeding on plant parts; %)	Time allocation of first invertebrate feeding bouts (before feeding on plant parts; %)
March (*n*=12 days)	100.0	0.0
April (*n*=12 days)	100.0	0.0
May (*n*=12 days)	91.7	8.3
June (*n*=12 days)	100.0	0.0
July (*n*=12 days)	91.7	8.3
August (*n*=12 days)	91.7	8.3
October (*n*=10 days)	90.0	10.0
December (*n*=12 days)	83.3	16.7
Average	93.5	6.5
SD (±)	6.0	6.0
Min.	83.3	0.0
Max.	100.0	16.7

Table 4.7. Time allocation of first invertebrate feeding bouts by the La Yunai group during 2006.

Scientific name	%	Family	Plant part consumed
Dipteryx panamensis	14.0	Fabaceae/Papilonaceae	Embryo/Endosperm
Ficus americana [2]	11.9	Moraceae	Mesocarp/Seed
Bactris gasipaes [1][2]	7.4	Arecaceae	Coconut apex + Seed/ Coconut apex + Petal
Ficus insipida	6.8	Moraceae	Mesocarp/Seed
Inga spectabilis [2]	6.6	Fabaceae/Mimosaceae	Endocarp
Nephelium lappaceum [1][2]	6.0	Sapindaceae	Endocarp
Psidium guajava [1]	4.7	Myrtaceae	Mesocarp/Seed
Miconia affinis	4.1	Melastomataceae	Mesocarp/Seed
Alchornea costaricensis [2]	3.8	Euphorbiaceae	Mesocarp/Seed
Brosimum alicastrum	3.4	Moraceae	Endocarp
Inga marginata	3.1	Fabaceae/Mimosaceae	Endocarp/seed
Sterculia recordiana [2]	2.2	Sterculiaceae	Endocarp
Ficus tonduzii [2]	2.1	Moraceae	Mesocarp/Seed
Inga sapindoides	1.6	Fabaceae/Mimosaceae	Endocarp
Apeiba membranacea	1.5	Tiliaceae	Mesocarp/Seed
Inga ruziana [2]	1.4	Fabaceae/Mimosaceae	Endocarp
Cestrum megalophylum [2]	1.3	Solanaceae	Mesocarp/Seed
Ficus pertusa	1.3	Moraceae	Mesocarp/Seed
Passiflora ambigua	1.0	Passifloraceae (liana)	Mesocarp/Seed
Inga tonduzii	0.96	Fabaceae/Mimosaceae	Endocarp
Ficus spp. 2 [2(?)]	0.83	Moraceae	Mesocarp/Seed
Inga densiflora [2]	0.76	Fabaceae/Mimosaceae	Endocarp/seed
Dendropanax arboreus	0.60	Araliaceae	Endocarp
Sapium grandulosum [2]	0.60	Euphorbiaceae	Mesocarp/Seed
Pouteria spp [2(?)]	0.56	Sapotaceae	Endocarp
Virola koschnyi [2]	0.53	Myristicaceae	Endocarp
Calatola costaricensis [2]	0.51	Icacinaceae	Mesocarp
Artocarpus altilis	0.48	Moraceae	Petiole
Rhodostemonodaphne kunthiana [2]	0.48	Lauraceae	Endocarp/seed

Table 4.8. Plant species consumed by the La Yunai group during 2006.

Scientific name	%	Family	Plant part consumed
Hieronyma alchorneoides	0.47	Euphorbiaceae	Mesocarp/Seed
Cananga odorata [2]	0.45	Annonaceae	Mesocarp
Minquartia guianensis [2]	0.44	Olacaceae	Endocarp
Piper sancti-felicis [2]	0.44	Piperaceae	Mesocarp/Seed
Byrsonima crispa [2]	0.43	Malpighiaceae	Mesocarp/Seed
Ficus spp. 1 [2(?)]	0.41	Moraceae	Mesocarp/Seed
Ficus tricifolia [2]	0.40	Moraceae	Mesocarp/Seed
Virola sebifera	0.37	Myristicaceae	Endocarp
Duguetia panamensis [2]	0.36	Annonaceae	Mesocarp/Seed
Cocos nucifera [1]	0.35	Arecaceae	Coconut apex + Mesocarp[3]
Coussapoa villosa [2]	0.33	Cecropiaceae	Endocarp
Castilla elastica	0.32	Moraceae	Mesocarp
Laetia procera	0.32	Flacourtiaceae	Endocarp
Urera baccifera [2]	0.32	Urticaceae	Mesocarp/Seed
Poulsenia armata	0.31	Moraceae	Endocarp
Cordia lucidula [2]	0.28	Boraginaceae	Endocarp
Chrysophyllum cainito	0.24	Sapotaceae	Mesocarp/Seed
Conostegia xalapensis	0.24	Melastomataceae	Mesocarp/Seed
Drypetes brownii [2]	0.23	Euphorbiaceae	Mesocarp/Seed
Ficus nymphaeifolia [2]	0.23	Moraceae	Mesocarp/Seed
Pouteria torta	0.23	Sapotaceae	Endocarp + Seed
Otoba novogranatensis [2]	0.21	Myristicaceae	Endocarp
Trophis racemosa	0.21	Moraceae	Mesocarp/Seed
Protium spp 1 [2(?)]	0.20	Burseraceae	Mesocarp/Seed
Ficus spp (Nov Sp.[?]) [2(?)]	0.19	Moraceae	Mesocarp/Seed
Inga oerstediana [2]	0.17	Fabaceae/Mimosaceae	Endocarp/seed
Compsoneura sprucei [2]	0.16	Myristicaceae	Endocarp
Guatteria lucens [2]	0.16	Annonaceae	Mesocarp/Seed
Theobroma cacao [1]	0.16	Sterculiaceae	Mesocarp
Cordia bicolor	0.15	Boraginaceae	Endocarp
Pourouma minor	0.15	Cecropiaceae	Endocarp
Cordia cymosa [2]	0.12	Boraginaceae	Endocarp
Socratea exorrhiza	0.09	Arecaceae	Coconut apex
Cecropia obtusifolia [2]	0.07	Cecropiaceae	Petal
Maripa nicaraguensis [2]	0.07	Convolvulaceae (woudy liana)	Endocarp
Clusia spp [2(?)]	0.05	Clusiaceae (woudy epiphyte)	Mesocarp/Seed
Protium spp 2 [2(?)]	0.05	Burseraceae	Mesocarp/Seed
Cecropia insignis	0.04	Cecropiaceae	Petal
Hampea appendiculata [2]	0.04	Malvaceae	Mesocarp/Seed
Annona muricata [1]	0.01	Annonaceae	Mesocarp/Seed

Notes: [1] Introduced plant species; [2] New plant species report for the diet of capuchin monkeys (*Cebus* spp.; after Fragaszy *et al.* 2004. The table indicates that 47.8% (33 out of 69) of the plant species consumed by the La Yunai group have not been previously reported for the genus *Cebus* spp. (see Fragaszy *et al.* 2004). [3] In the case of coconuts (*Cocos nucifera*), the term mesocarp is used for the "coconut meat;" however, in this particular case, the endosperm actually is the appropriate term.
Eaten plant part: petiole (leaf shoot), petal (flowers), and fruits (and their annotated parts: seed [whole seed], mesocarp [flesh], Embryo/Endosperm [seed kernel], Endocarp [seed cover], Mesocarp [flesh], coconut apex [fleshy part around the contact area of the coconut with the palm stem]); 2) *n*=69 tree species, *n*=7262 observation bouts on feeding behavior.

Table 4.8. continued.

Family	%
Moraceae	28.9
Fabaceae/Mimosaceae	14.7
Fabaceae/Papilonaceae	14.0
Arecaceae	7.9
Sapindaceae	6.0
Euphorbiaceae	5.1
Myrtaceae	4.7
Melastomataceae	4.4
Sterculiaceae	2.4
Tiliaceae	1.5
Solanaceae	1.3
Myristicaceae	1.3
Passifloraceae (liana)	1.0
Sapotaceae	1.0
Annonaceae	0.99

Family	%
Araliaceae	0.60
Cecropiaceae	0.59
Boraginaceae	0.55
Icacinaceae	0.51
Lauraceae	0.48
Olacaceae	0.44
Piperaceae	0.44
Malpighiaceae	0.43
Flacourtiaceae	0.32
Urticaceae	0.32
Burseraceae	0.25
Convolvulaceae	0.07
Clusiaceae	0.05
Malvaceae	0.04

Note: n=29 tree families, n=7262 observation bouts on feeding behavior.

Table 4.9. Plant families consumed by the La Yunai group during 2006.

Species	# Individuals	% Individuals	Family
Bactris gasipaes	22	9.0	Arecaceae
Ficus americana	15	6.2	Moraceae
Miconia affinis	15	6.2	Melastomataceae
Inga sapindoides	13	5.3	Fabaceae/Mimosaceae
Psidium guajava	12	4.9	Myrtaceae
Alchornea costaricensis	8	3.3	Euphorbiaceae
Dipteryx panamensis	8	3.3	Fabaceae/Papilonaceae
Cananga odorata	7	2.9	Annonaceae
Inga marginata	7	2.9	Fabaceae/Mimosaceae
Cestrum megalophylum	6	2.5	Solanaceae
Ficus insipida	6	2.5	Moraceae
Inga spectabilis	6	2.5	Fabaceae/Mimosaceae
Brosimum alicastrum	5	2.1	Moraceae
Virola sebifera	5	2.1	Myristicaceae
Apeiba membranacea	4	1.6	Tiliaceae
Calatola costaricensis	4	1.6	Icacinaceae
Ficus pertusa	4	1.6	Moraceae
Laetia procera	4	1.6	Flacourtiaceae
Passiflora ambigua	4	1.6	Passifloraceae (liana)
Piper sancti-felicis	4	1.6	Piperaceae
Cocos nucifera	3	1.2	Arecaceae
Conostegia xalapensis	3	1.2	Melastomataceae
Dendropanax arboreus	3	1.2	Araliaceae
Inga ruziana	3	1.2	Fabaceae/Mimosaceae
Nephelium lappaceum	3	1.2	Sapindaceae
Pouteria torta	3	1.2	Sapotaceae
Sapium grandulosum	3	1.2	Euphorbiaceae
Socratea exorrhiza	3	1.2	Arecaceae
Sterculia recordiana	3	1.2	Sterculiaceae
Trophis racemosa	3	1.2	Moraceae
Virola koschnyi	3	1.2	Myristicaceae

Table 4.10. Number of individual trees used for feeding by the La Yunai group during 2006.

Species	# Individuals	% Individuals	Family
Byrsonima crispa	2	0.82	Malpighiaceae
Castilla elastica	2	0.82	Moraceae
Cordia bicolor	2	0.82	Boraginaceae
Cordia cymosa	2	0.82	Boraginaceae
Cordia lucidula	2	0.82	Boraginaceae
Drypetes brownii	2	0.82	Euphorbiaceae
Duguetia panamensis	2	0.82	Annonaceae
Ficus spp. 2	2	0.82	Moraceae
Ficus tonduzii	2	0.82	Moraceae
Inga densiflora	2	0.82	Fabaceae/Mimosaceae
Inga oerstediana	2	0.82	Fabaceae/Mimosaceae
Minquartia guianensis	2	0.82	Olacaceae
Pouteria sp	2	0.82	Sapotaceae
Rhodostemonodaphne kunthiana	2	0.82	Lauraceae
Annona muricata	1	0.41	Annonaceae
Artocarpus altilis	1	0.41	Moraceae
Cecropia insignis	1	0.41	Cecropiaceae
Cecropia obtusifolia	1	0.41	Cecropiaceae
Chrysophyllum cainito	1	0.41	Sapotaceae
Clusia sp.	1	0.41	Clusiaceae
Compsoneura sprucei	1	0.41	Myristicaceae
Coussapoa villosa	1	0.41	Cecropiaceae
Ficus nymphaeifolia	1	0.41	Moraceae
Ficus sp. (Nov Sp.[?])	1	0.41	Moraceae
Ficus spp. 1	1	0.41	Moraceae
Ficus tricifolia	1	0.41	Moraceae
Guatteria lucens	1	0.41	Annonaceae
Hampea appendiculata	1	0.41	Malvaceae
Hieronyma alchorneoides	1	0.41	Euphorbiaceae
Inga tonduzii	1	0.41	Fabaceae/Mimosaceae
Maripa nicaraguensis	1	0.41	Convolvulaceae
Otoba novogranatensis	1	0.41	Myristicaceae
Poulsenia armata	1	0.41	Moraceae
Pourouma minor	1	0.41	Cecropiaceae
Protium spp. 1	1	0.41	Burseraceae
Protium spp. 2	1	0.41	Burseraceae
Theobroma cacao	1	0.41	Sterculiaceae
Urera baccifera	1	0.41	Urticaceae

Note: n=244.

Table 4.10. continued.

	1st	2nd	3rd	Total %
March (n=1181)	*Dipteryx panamensis*: 46.3% (88.9%; 8/9)	*Inga spectabilis*: 14.9% (33.3%; 2/6)	*Sterculia recordiana*: 13.9% (75%; 2/3)	75.1
April (n=750)	*Dipteryx panamensis*: 38.9% (44.4%; 4/9)	*Inga spectabilis*: 17.4% (66.7; 4/6)	*Ficus americana*: 16.4% (20%; 3/15)	72.7
May (n=755)	Miconia affinis: 24.6% (66.7; 10/15)	*Ficus americana*: 17.5% (20%; 3/15)	*Dipteryx panamensis*: 14.7% (22.2%; 2/9)	56.8
June (n=940)	*Ficus americana*: 38.1% (20%; 3/15)	Brosimum alicastrum: 10.5% (60.7%; 3/5)	*Ficus pertusa*: 9.05% (50%; 2/4)	57.7
July (n=399)	*Inga spectabilis*: 29.9% (33.3%; 2/6)	Miconia affinis: 15.7% (46.7%; 7/15)	Piper sancti-felicis: 8.2% (100%; 4/4)	53.8
August (n=833)	*Ficus insipida*: 60.4% (83.3%; 5/6)	Psidium guajava : 11.3% (75%; 9/12)	*Bactris gasipaes*: 6.8% (31.8%; 7/22)	78.5
September (Exp.1) (n=1001)	*Nephelium lappaceum*: 15.8% (66.7%; 2/3)	*Ficus americana*: 10.5% (13.3%; 2/15)	Psidium guajava: 10.2% (16.7; 2/12)	36.5
October (n=274)	*Nephelium lappaceum*: 31.8% (66.7; 2/3)	*Bactris gasipaes*: 20.3% (40.9%; 9/22)	*Dipteryx panamensis* : 12.8% (11.1%; 1/9)	64.9
October (Exp. 2) (n=775)	*Nephelium lappaceum*: 32.9% (66.7; 2/3)	*Bactris gasipaes*: 27.8% (31.8%; 7/22)	Brosimum alicastrum: 9.7% (40%; 2/5)	70.4
November (Exp. 2) (n=660)	*Bactris gasipaes*: 25.2% (9.1%; 2/22)	Inga marginata: 19.3% (28.6%; 2/7)	Alchornea costaricensis: 14.4% (87.5; 7/8)	58.9
December (n=253)	Alchornea costaricensis: 55.1% (25%; 2/8)	Inga ruziana: 22.9% (66.7%; 2/3)	Cocos nucifera: 7.8% (100%; 3/3)	85.8

Note: In bold, tree species that coincided with the most used tree families that were uitilized as feeding sources by the capuchins.
Note. n=7262 observation bouts on feeding behavior. In parentheses, number of feeding trees used/Total number of individual feeding trees of the same given species, and proportion (percentages) of the feeding trees used *vs.* the total number of individual feeding trees of the same given species.

Table 4.11. Top three feeding tree species used by the La Yunai group per month during 2006.

Table 4.12. Plant items and parts of plant species consumed by the La Yunai group during 2006.

Table 4.13. Top three preferred parts of plant species consumed by the La Yunai group per month during 2006.

Plant item	Plant eaten part	%
Fruit	Mesocarp/Seed	44.4
Fruit	Endocarp	29.1
Fruit	Embryo/Endosperm	14.0
Fruit	Endocarp/Coconut apex	6.6
Fruit	Endocarp/Seed	2.6
Fruit	Mesocarp	1.8
Fruit	Coconut apex	0.4
Flower	Petal	0.7
Leave shoot	Petiole	0.5

Note: n=7262 observation bouts on feeding behavior.

	1st	2nd	3rd
March (n=1181)	Embryo/Endosperm (46.3)	Endocarp (38.0)	Mesocarp/Seed (9.5)
April (n=750)	Embryo/Endosperm (38.9)	Mesocarp/Seed (36.9)	Endocarp (19.4)
May (n=755)	Mesocarp/Seed (63.2)	Endocarp/Seed (19.3)	Embryo/Endosperm (14.7)
June (n=927)	Mesocarp/Seed (69.4)	Endocarp (23.0)	Embryo/Endosperm (5.8)
July (n=399)	Mesocarp/Seed (47.9)	Endocarp (40.2)	Mesocarp (6.0)
August (n=833)	Mesocarp/Seed (80.7)	Endocarp (12.0)	Endocarp/Coconut apex (5.8)
September (Exp.1) (n=1001)	Endocarp (42.8)	Mesocarp/Seed (40.2)	Mesocarp (6.6)
October (n=274)	Endocarp (52.4)	Endocarp/Coconut apex (20.3)	Embryo/Endosperm (12.8)
October (Exp. 2) (n=775)	Endocarp (52.2)	Endocarp/Coconut apex (27.8)	Mesocarp/Seed (17.2)
November (Exp. 2) (n=660)	Mesocarp/Seed (44.7)	Endocarp (28.0)	Endocarp/Coconut apex (25.2)
December (n=253)	Mesocarp/Seed (66.9)	Endocarp (25.3)	Mesocarp (7,8)

Note: n=7262 observation bouts on feeding behavior. Numbers in parentheses are percentages.

Scientific name	%	Family		Scientific name	%	Family
Pentaclethra macroloba	20.3	Fabaceae/Mimosaceae		Zygia longifolia	0.97	Fabaceae/Mimosaceae
Apeiba membranacea	13.8	Tiliaceae		Bravaisia integerrina	0.77	Acanthaceae
Stryphnodendron microstachyum	10.8	Fabaceae/Mimosaceae		Cocos nucifera [1]	0.77	Arecaceae
Erythrina poeppigiana [1]	7.4	Fabaceae/Papilonaceae		Tapirira guianensis	0.77	Anacardiaceae
Ficus insipida	5.6	Moraceae		Brosimum alicastrum	0.73	Moraceae
Alchornea costaricensis	4.5	Euphorbiaceae		Terminalia bucidoides	0.70	Combretaceae
Guatteria aeruginosa	4.4	Annonaceae		Hampea appendiculata	0.67	Malvaceae
Ficus pertusa	3.4	Moraceae		Tectona grandis [1]	0.67	Verbenaceae
Nephelium lappaceum [1]	3.4	Sapindaceae		Virola koschnyi	0.50	Myristicaceae
Guatteria lucens	3.3	Annonaceae		Ficus spp. 1	0.33	Moraceae
Psidium guajava [1]	2.7	Myrtaceae		Dipteryx panamensis	0.30	Fabaceae/Papilonaceae
Pouteria sp.	2.6	Sapotaceae		Ceiba petandra	0.23	Bombaceae
Virola sebifera	2.4	Myrtaceae		Calatola costaricensis	0.20	Icacinaceae
Hieronyma alchorneoides	1.8	Euphorbiaceae		Chrysophyllum venezuelense	0.17	Sapotaceae
Goethalsia meiantha	1.8	Tiliaceae				
Ficus americana	1.5	Moraceae		Cananga odorata	0.10	Annonaceae
Dendropanax arboreus	1.3	Araliaceae		Ochroma pyramidale	0.07	Bombaceae
Luehea seemannii	1.2	Tiliaceae				

Note: [1] Introduced plant species; n=34 tree species; n=3858 observation bouts

Table 4.14. Plant species used as resting trees by the La Yunai group during 2006.

Resting sites

Pentaclethra macroloba was the most common canopy tree species in the study group's home range (Chapter 3: Table 3.2) and was the most commonly used resting tree for members of the La Yunai group. Twenty-point-three percent of capuchin resting time occurred in *P. macroloba* (Tables 4.15, 4.16, 4.17). At EBLS, individual trees of this species had a mean DBH of 48.3±30.1, a mean height of 20.1±4.5 m, and a mean crown diameter of 13.0±3.7 m. Two other tree species (*Apeiba membranacea* and *Stryphnodendron microstachyum*) accounted for 24.6% of resting time (Table 4.14). *A. membranacea* is relatively common (10 individuals/hectare) based on the floristic composition of the sample transects, but *S. microstachyum* was less common (<5 ind./ha) (Chapter 3: Table 3.2). In all study months (except December), these three species were the most frequently used resting trees (in terms of time spent resting; see above, and Table 4.17). In general, resting trees exhibited a mean DBH of 67.2±57.0, a mean tree height of 20.6±6.1 m, and a mean crown diameter of 12.5±4.6 m.

Species	# Individuals	% Individuals	Family
Pentaclethra macroloba	17	22.7	Fabaceae/Mimosaceae
Erythrina poeppigiana	4	5.3	Fabaceae/Papilonaceae
Ficus insipida	4	5.3	Moraceae
Alchornea costaricensis	3	4.0	Euphorbiaceae
Hieronyma alchorneoides	3	4.0	Euphorbiaceae
Virola sebifera	3	4.0	Myrtaceae
Apeiba membranacea	2	2.7	Tiliaceae
Brosimum alicastrum	2	2.7	Moraceae
Cocos nucifera	2	2.7	Arecaceae
Dendropanax arboreus	2	2.7	Araliaceae
Ficus americana	2	2.7	Moraceae
Ficus pertusa	2	2.7	Moraceae
Goethalsia meiantha	2	2.7	Tiliaceae
Guatteria aeruginosa	2	2.7	Annonaceae
Luehea seemannii	2	2.7	Tiliaceae
Pouteria sp.	2	2.7	Sapotaceae
Psidium guajava	2	2.7	Myrtaceae
Tectona grandis	2	2.7	Verbenaceae
Zygia longifolia	2	2.7	Fabaceae/Mimosaceae
Bravaisia integerrina	1	1.3	acanthaceae
Calatola costaricensis	1	1.3	Icacinaceae
Cananga odorata	1	1.3	Annonaceae
Ceiba petandra	1	1.3	Bombaceae
Chrysophyllum venezuelense	1	1.3	Sapotaceae
Dipteryx panamensis	1	1.3	Fabaceae/Papilonaceae
Ficus spp. 1	1	1.3	Moraceae
Guatteria lucens	1	1.3	Annonaceae
Hampea appendiculata	1	1.3	Malvaceae
Nephelium lappaceum	1	1.3	Sapindaceae
Ochroma pyramidale	1	1.3	Bombaceae
Stryphnodendron microstachyum	1	1.3	Fabaceae/Mimosaceae
Tapirira guianensis	1	1.3	Anacardiaceae
Terminalia bucidoides	1	1.3	Combretaceae
Virola koschnyi	1	1.3	Myristicaceae

Note: n=75

Family	%
Fabaceae/Mimosaceae	32.0
Tiliaceae	16.8
Moraceae	11.6
Annonaceae	7.8
Fabaceae/Papilonaceae	7.7
Euphorbiaceae	6.3
Myrtaceae	5.1
Sapindaceae	3.4
Sapotaceae	2.7
Araliaceae	1.3
Acanthaceae	0.77
Anacardiaceae	0.77
Arecaceae	0.77
Combretaceae	0.70
Malvaceae	0.67
Verbenaceae	0.67
Myristicaceae	0.50
Bombaceae	0.30
Icacinaceae	0.20

Note: n=33 tree families

Table 4.15. Number of individual trees used for resting by the La Yunai group during 2006.

Table 4.16. Plant families used as resting trees by the La Yunai group during 2006.

	1st	2nd	3rd
March (n=167)	*Stryphnodendron microstachyum*: 40.8% (1/1; 100%)	*Pentaclethra macroloba*: 26.5% (2/17; 11.8%)	*Hieronyma alchorneoides*: 17.0% (25%; 1/4)
April (n=311)	*Pentaclethra macroloba*: 86.96% (: 29.4%; 5/17)	*Hieronyma alchorneoides*: 8.7% (25%; 1/4)	*Stryphnodendron microstachyum*: 4.35% (100%; 1/1)
May (n=497)	*Stryphnodendron microstachyum*: 32.9% (100%, 1/1)	*Pentaclethra macroloba*: 20.5% (23.5%; 4/17)	*Virola sebifera*:15.6% (100%; 3/3)
June (n=365)	*Alchornea costaricensis*: 26.34% (33.3%, 1/3)	*Ficus americana*: 24.7% (100%, 2/2)	*Pentaclethra macroloba*: 15.38% (5.9%, 1/17)
July (n=200)	*Pentaclethra macroloba*: 47.3% (5.9%; 1/17)	*Guatteria aeruginosa*: 14.2% (100%; 2/2)	*Tapirira guianensis*: 13.6% (100%; 1/1)
August (n=218)	*Erythrina poeppigiana*: 65.6% (50%; 2/4)	*Ficus insipida*: 27.6% (100%; 4/4)	*Hampea appendiculata* : 2.7% (100%; 1/1)
September (Exp. 1) (n=763)	*Pentaclethra macroloba*: 24.8% (41.2%; 7/17)	*Ficus pertusa*: 14.0% (33.3%; 1/3)	*Stryphnodendron microstachyum*: 13.8% (100%, 1/1)
October (n=336)	*Nephelium lappaceum*: 20.2% (66.7%; 2/3)	*Apeiba membranacea*: 18.3% (100%; 2/2)	*Pentaclethra macroloba*: 15.7% (17.6%; 3/17); *Guatteria aeruginosa*: 15.7% (50%; 1/2)
October (Exp. 2) (n=678)	*Guatteria lucens*: 19.4% (50%; 1/2)	*Alchornea costaricensis*: 14.3% (100%; 3/3)	*Ficus insipida*: 9.6% (100%; 4/4)
November (Exp. 2) (n=270)	*Apeiba membranacea*: 59.6% (50%; 1/2)	*Erythrina poeppigiana*: 13.7% (50%; 2/4)	*Pentaclethra macroloba*: 10.1% (17.6%; 3/17)
December (n=102)	*Cocos nucifera*: 27.4% (75%; 3/4)	*Guatteria aeruginosa*: 17.9% (50%; 1/2)	*Goethalsia meiantha*: 15.5% (100%; 1/1)

Note: In bold, tree species that coincided with the most used tree families that were utilized as resting sites by the capuchins. In parentheses, number of used resting trees/Total number of individual resting trees of the same given species, and proportion (percentages) of the used resting trees *vs.* the total number of individual resting trees of the same given species. n=3858 observation bouts.

Table 4.17. Top three resting tree species used by the La Yunai group per month during 2006.

Ranging patterns

Based on data collected during 86 days, the home range of the La Yunai group was 45.5 ha. The home range was calculated from the cumulative maps of day range during the study period. This area includes 22.5 ha within the EBLS and 22.9 ha outside the field station (Table 4.18). Overall, the home range area of the La Yunai group is similar to that reported for *C. imitator* at three sites across its range (Table 4.23). The home range of the La Yunai group overlapped minimally (0.5 ha at the northern tip of the home range) with one group of capuchins that inhabited the adjacent "large forest." This area represents 1.1% of the total home range. A neighboring northern capuchin group came into contact on two days during the study period. The estimated group density for the La Yunai group was 26.4 individuals/km^2, which is similar to that reported for *C. imitator* at other sites (Table 4.23).

Based on 86 complete follows, the estimated daily path length varied from 1980±450 m in October to 2880±658m in May, with a mean of 2339±626 m during the study period (Table 4.19). Monthly variation in day range was not significantly different (χ^2=484.94 d.f.=7, p<0.01). Areas covered by primary and advanced secondary forests were mainly used by the capuchins, occupying 66.8% of the total home range.

Type of vegetation	Abbreviation	Within EBLS Ha	%	Outside EBLS Ha	%	Total Ha	%
Advanced secondary forest	ASF	6.0	27.6	5.3	23.0	11.8	25.4
Bambuzal	B	0.64	2.7	0.40	1.7	1.0	2.2
Charral	Ch	1.2	5.2	0.27	1.2	1.5	3.2
Platanillal	Pl	0.30	1.3	-	-	0.30	0.65
Primary forest	PF	8.4	36.2	10.8	47.0	19.4	41.5
Secondary forest	SF	1.1	4.5	0.74	3.2	1.8	3.9
Tacotal or schrub	T	5.1	22.6	3.2	14.1	8.6	18.4
Living fence	LF	-	-	1.5	6.5	1.5	3.2
Plantation	P	-	-	0.76	3.3	0.76	1.6
Total		22.5		22.9		45.5	

Table 4.18. Home range and types of vegetation used by the La Yunai group during 2006.

	March	April	May	June	July	August	October	Decemb.	Total
Average (m)	2345	2252	2880	2548	2244	2006	1980	2365	2328
SD (±)	562	476	658	451	577	727	450	794	626
Min. (m)	1060	1518	2013	1895	1303	923	1047	1431	923
Max. (m)	3266	3307	4391	3214	2985	3265	2568	3431	4391

Table 4.19. Daily range of the La Yunai group during 2006.

Discussion

Changes in food availability and distribution during different months of the year have been argued to play an important role in primate diet, foraging strategies, social interactions, reproduction, and activity budget (Clutton-Brock and Harvey 1977, Garber 1987, van Schaik and Brockman 2005 and papers herein). Terborgh (1986) argued that particularly under conditions of food scarcity, vertebrate consumers rely on a reduced set of plant species that produce a large food reward over an extended fruiting period. These resources are named keystone plant resources (KPRs). KPRs are assumed to represent relatively low-density tree species that are constantly exploited from year-to-year by different vertebrate taxa of a given forest community. Marshall and Wrangham (2007) present an alternative concept to explain primate strategies during food-limited periods of the year. They argue that primate foragers exploit low-quality resources such as bark or mature leaves when commonly used resources are not available. These resources are called fallback foods (FBFs). Alternatively, Sussman (1987) suggested that although primates may exploit >50 plant species, they primarily rely on a small number of plant species throughout the year. These resources (≤10 plant species) accounted for ≥50% of plant feeding time, and are here termed as highly selected foods (HSFs) (see Table 4.20).

In this chapter, I examine the degree to which white-faced capuchins feeding behavior is most consistent with the use of KPRs, FBFs or HSFs. Table 4.20 summarizes these concepts and the specific factors that served to distinguish KPRs from FBFs to HSFs.

Category	KPRs	FBFs	HSFs
Vertebrate taxa involved	Birds and mammals (incl. i.e. primates and ungulates)	Primate-specific. Probably valid for other vertebrate taxa	Primate-specific. Probably valid for other vertebrate taxa
Timing of use	Certain months of the year, principally the dry season	Limited periods of the year, principally the dry season	Year-round
Type of resource	Limited number of plant species exploited by the entire vertebrate community	Small number of hard-to-process resources used by only one or principally one primate species in the community	Small subset of different plant resources per month and per year
Resource exploitation regime	Consistent, used year after year	Consistent, used year after year	Consistent use of ≤10 plant species per month and per year
Feeding time	Used heavily, major resource	From reduced percentage to up to 100%	≥50% (for a small set of plant species)
Quality of resource	Not specified	Low	Low or high
Feeding adaptation	Yes	Yes	Not necessarily
Abundance	Rare	Abundant	Variable, site-dependent

Table 4.20. Contrasting the concepts of keystone plant resources (KPRs), fallback foods (FBFs), and highly selected food (HSFs).

Activity budget

Capuchins at EBLS were found to exploit a highly consistent activity budget (Table 4.2) throughout the study year. Stability was defined as a pattern of consistent and similar relative frequency of a given behavioral repertoire across different periods of the year (Garber 1993). As reported in other studies on *C. imitator*, the La Yunai group spends 20% of the activity budget on feeding. White-faced capuchins living in other wet forest sites (Barro Colorado Island-Panama and Trujillo-Honduras) and white-faced capuchins living in dry forests (Santa Rosa National Park-Costa Rica) also are characterized by activity budgets in which ~20% of the time is devoted to feeding (Table 4.21). Foraging (52.6%) was the most common activity reported in our study group.

Foraging was defined as the movement within or between single or multiple trees when searching for a food item (fruit, insect, seed) (see Chapter 3). Rose (1998) defined "total foraging" as the active search for food, or the sum of her categories "eat" plus "forage." She found that capuchins devoted ~15% of their activity budget to foraging, and 35% to eating. In contrast, Mitchell (1989) who defined foraging as the active search for food reported that on Barro Colorado Island, a rainforest in Panama, white-faced capuchins spent 57% of their activity budget foraging. Although several factors may account for the differences reported across field sites, contrasting definitions of the behavioral category of "foraging" is likely to be a major factor.

Diet, feeding, and foraging behavior

The La Yunai group members spent 57% of their feeding and foraging time on plant parts and 42.9% on animal matter (Table 4.5). These values were similar across the study period and were not statistically different between months. For example, in March 59.7% of the time was allocated to feeding/foraging on plant parts and 40.4% to animal matter. In August, 61.8% of the time was used in feeding and foraging on plant parts and 38.2% on animal matter. Thus, the pattern of time devoted to feeding and foraging was constant and varied little throughout the year.

In contrast, in two dry forests in northwestern Costa Rica, white-faced capuchin feeding and foraging time used for obtaining plant parts and animal matter were more variable during certain periods of the year. For example, at Santa Rosa, the exploitation of plant matter, mainly seeds and mesocarp, varied from 82.7% of feeding time in the wet season (June to November) to a low of 43.4% in the dry season (December to May) for a single capuchin group (Rose 1998). At Curú, a dry forest site in northwestern Costa Rica, Baker (1998) reported that white-faced capuchins allocated 56% and 41% of their feeding time to plant parts and animal prey, respectively. Overall, in response to seasonal changes in food availability, white-faced capuchins switch to a diet based either on plant parts or animal matter; however, the time devoted to feeding and foraging remains relatively stable (Table 4.21, Table 4.22).

The members of the La Yunai group were found to feed mainly on *Dipteryx panamensis* (14.0% of feeding time). *D. panamensis* is an asynchronously fruiting tree species that according to Chun (2008) serves as a KPR in the Neotropics. Chun (2008) based this assessment on the fact that this tree species fruits when most other tree species are not producing ripe fruits. *D. panamensis* produces a single capsule with one seed. Most of the protein and lipids of the fruit of *Dipteryx* are found in the seed (Betancourt and Torres, 1999). Betancourt *et al.* (1999) reported that *Dipteryx punctata* contains 6.4% proteins and 7% lipids. These nutrients are concentrated mainly in the seed (embryo) which is the part consumed by the capuchins at EBLS during this study. The seed is covered by a fiber-rich endocarp/pericarp (21.1% of fibrous material in *D. punctata*, after Betancourt *et al.* 1999) which is discarded by the capuchins. *D. panamensis* was consumed as one of the top three feeding species during four months of the 8-month study period. *D. panamensis* is a large emergent tree with a large crown volume. Each tree may produce a mast fruit crop with up to 10,000 fruits per year (Bonaccorso *et al.* 1980). Fruit production of this genus is asynchronous at EBLS, and tree individuals of this species are among the largest canopy trees in this forest. The reliance of capuchins at EBLS on this tree species has important conservation implications because it is used commercially for its hardwood (Aguilar 2007: Pers. comm.). For that reason, *D. panamensis* has been declared as a tree vulnerable to extinction according to Costa Rican law.

This study troop relied on *Ficus* spp. (Moraceae) and *Inga* spp. (Fabaceae/Mimosaceae) during seven months of the study period. Figs also are asynchronous fruiting tree species and produce a large mast crop. The fruit is small and composed of a soft pulp which is highly nutritious (Duke and Atchley 1986, Conklin and Wrangham 1994, O′Brian et al. 1998). For example, *Ficus insipida*, the second most common fig species eaten by the capuchins at EBLS is reported to contain 7.4% crude protein and 4.4% lipids (calculated from data in Hladik *et al.* 1971, Milton *et al.* 1980, Serio-Silva *et al.* 2002). Terborgh (1983) reported the high reliance of figs by *S. apella* and *C. albifrons* in Peru. Figs were rare within the home ranges of these two capuchin species. At EBLS, figs also were rare (Chapter 3: Table 3.2). Depending on the species, in our forest sample, fig tree species had a density of ≤0.64 individuals/ha.

A major limitation of concepts such as KPRs and FBFs is that they fail to account for the considerable local variability in plant production and specific plant species consumed by a group of primates over a span of several years. For example, based on a 20-year study of howler monkeys on Barro Colorado Island-Panama, Milton (2005) reports that fluctuation in fruit production on individual trees and tree species in any single year or between years was not closely correlated with abiotic factors such as rainfall. According to Milton (2005), the variability in the yearly production of individual tree species makes difficult to predict how different vertebrate taxa are affected by fruit availability.

Figs are consumed by virtually all primates in forests where they are found. Fig fruits also contain insect larvae which serve to increase their nutritional value (Urquiza-Haas *et al.* 2008). Moreover, although figs do not require any specific anatomical adaptation for their ingestion, efficient exploitation across different months of the year may be dependent on cognitive abilities associated with locating and relocating fig-bearing trees in the forest. At EBLS, figs are best described as HSFs due to the fact they produce a large fruit crop, are simple to process, are relatively nutritious, and are among the top three most consumed tree species during four months of the year.

Nuts of *Bactris gasipaes* (Arecaceae) were mainly ingested by white-faced capuchins for three consecutive months. During this period other major tree species exploited by white-faced capuchins were *Nephelium lappaceum* and *Dipteryx panamensis*. Synchrony in fruit production has been reported for *B. gasipaes* in other Mesoamerican rainforests such as Barro Colorado Island in Panama (De Steven *et al.* 1987). This palm produces energetically-rich fruits. For each 100 g. of pulp, 37 g. are carbohydrates, 4.7 g. lipids, and 3.8 g. protein (SIDBAAP 2009). In addition, 100 g. of *B. gasipaes* (each fruit weight 28.5 g. [Soto *et al.* 2005]) provides up to 194 calories (SIDBAAP 2009). At EBLS, capuchins consumed the oily-rich and "butter-like" part of the fruit which is located in the apex. The common use of palm nuts also has been reported in *S. apella* and *C. albifrons* living in rainforests in Amazonia (Terborgh 1983, Spironello 1991, Peres 1994). Two other tufted capuchin species (*S. libidinosus* and *S. xanthosternos*) also exploit palm nuts. These species are able to open the hard outer shell by using tools (Moura and Lee 2004, Visalberghi *et al.* 2004a, 2004b, Rodrigues-Canale *et al.* 2009).

Nut-cracking using tools has not been reported for any capuchin population living in rainforests (see review: Urbani and Garber 2002), including EBLS (Urbani this study). It has been argued that using tools for nut-cracking behavior is a response to food shortage and the use of fallback foods in a harsh environment ("energy bottleneck") (Moura and Lee 2004). In addition, it has been argued that extensive foraging on the ground is a precondition for tool-using behavior in capuchins (Visalberghi *et al.* 2004b). Considering the lack of nut-cracking data from EBLS and other localities where wild capuchins have been studied over several years, the high levels of nut-cracking by bearded capuchins living in semi-arid *Caatinga* forests in northeastern Brazil may be a behavioral response to foraging challenges encountered in an extremely dry deciduous habitat (rainfall ranges between 300 and 1000 mm per year, and the dry period lasts for ~8 months [Paganucci de Queiroz 2006], 75 to 100% of the trees lose their leaves for at least 6 months of the year [Bullock 1995, Machado *et al.* 1997, Paganucci de Queiroz 2006]). Under such conditions, capuchins are dependent on plant resources that are available on the ground including palm nuts (Visalberghi *et al.* 2004a) and succulent tubers (Moura and Lee 2004, Ottoni and Mannu 2008) (support for the "energy bottleneck" contention of Moura and Lee [2004], see above). On the other hand, *Caatinga* sites are found to have broken and unconnected canopies composed of a matrix of open shrubs and tree branches mainly near ground level (Lemos and Rodal 2002). In continuous canopy forests, capuchins can travel and forage throughout their home range arboreally, spending less time on the ground. In *Caatinga* forests, capuchins are reported to commonly feed, forage, and travel on the ground (support for the precondition contention of Visalberghi *et al.* [2004b], see above). Another factor that may help explain differences in capuchin foraging and tool use is that capuchin species such as *Sapajus libidinosus* and *S. xanthosternos* live in habitats dominated by palms and trees of the family Leguminosae (Paganucci de Queiroz 2006). Legumes are encapsulated fruits that are difficult to open. Some capuchin populations living in these dry habitats may have by chance or trial-and-error developed a behavioral pattern of using tools to open or access these resources. In sum, the combination of a harsh environment, terrestrial foraging, and the exploitation of difficult-to-open encapsulated foods may have resulted in the use of tools by bearded capuchins in northeastern Brazil.

At EBLS, Orthoptera, Hymenoptera, and Isoptera are the most commonly consumed invertebrates. From studies in Amazonian Peru, Garber (1989) suggested that orthopterans provide high-quality nutrients to tamarins. In an Amazonian site in north-central Brazil, Penny and Arias (1982) estimated that orthopterans represent a significant part of the total biomass in the lower forest strata. For example, using flight traps orthopterans reached 10% of the total biomass of insects collected. Twenty-one percent of the insect biomass collected in light traps located in the forest understory also were orthopterans. Based on a nutritional analysis of orthopterans ($n=149$) there is evidence that these invertebrates are rich in lipids and contain twice as much crude protein as plant parts commonly consumed by primates (Hernández 2001: Pers. comm., Harrison-Levine1 *et al.* 2003). At EBLS, white-faced capuchins rely on orthopterans as energy-rich food year-round.

As suggested by Robinson (1986) and Freese and Oppenheimer (1981) for wedge-capped capuchins, white-faced capuchins at EBLS consumed fruits in the early morning. Once light levels increase in the lower canopy, white-faced capuchins began to forage for more cryptic prey such as invertebrates. In the present study, the capuchins caught and ingested two squirrels, and on rare occasions were observed to consume lizards, birds, and bird eggs. In comparison, Rose (1997) reported that in a dry forest in northwestern Costa Rica, white-faced capuchins engaged in hunting vertebrates mainly squirrels and coati pups. She reported that hunting accounted for 5.4 prey per 100 hours, and included capuchin foragers acting alone or the coordinated activity of two or more individuals. At EBLS, both cases of successful squirrel hunting involved at least two individuals.

At EBLS, six plant species accounted for >50% of white-faced yearly feeding time. Some of these tree species fruit asynchronously, and their fruits were available during different months of the year. As in the case of plant parts, the consumption of invertebrates varied little across the study months. Thus, the foraging and feeding patterns of white-faced capuchins at EBLS are best described as stable year-round. In this regard the cognitive and behavioral abilities of white-faced capuchins enable them to consistently and mainly rely on a limited number of different tree species during the year.

Ranging patterns

The home range of the La Yunai group is consistent with that reported for *C. imitator* at other sites (Table 4.23). The density of white-faced capuchins as EBLS (26.4 individuals/ km²) also is consistent with that found for white-faced capuchin monkeys at other sites (Table 4.23). For example, in studies in an Atlantic rainforest (Barro Colorado, Panamá:

Mitchel [1989]) and a Pacific dry forest (Palo Verde, western Costa Rica: Moscow and Vaughan [1987]), the densities were 22.2 ind/km², and 24.6 ind/km², respectively.

The mean daily path length of capuchins at EBLS fits within the average found for *C. imitator* at other sites and the genus *Cebus* in general (Table 4.23). The daily path length at EBLS range varied from 2880 m to 1980 m. It was greatest in May and lowest in the month of October. In a more seasonal site in northwestern Costa Rica, Baker (1998) reported that there was no significant variation in white-faced capuchins' day range between the wet (1498 m) and dry (1547 m) seasons. In contrast, in seasonal deciduous Venezuelan dry forests (Robinson [1984, 1986], Miller [1986, 1988]) found that wedge-capped capuchins (*C. olivaceus*) altered their ranging patterns by significantly increasing (3580 m) daily path length in the dry season and decreasing (1046 m) daily path length during the wet season. In general, the day length range in the different sites where capuchins had been studied remained similar (Table 4.23).

A hypothesis that may explain the existence of consistency in daily path length monthly day range at EBLS is the presence of several synchronous and asynchronous fruiting tree species (e.g. *Dipteryx*, *Ficus*, *Bactris*, *Nephelium* and *Inga*) that serve as HSFs year-round. During periods in which one of these genera was not fruiting, at least one of the other genera was available and fruiting. Furthermore, the time devoted to invertebrate foraging remained relatively stable throughout the year. A similar pattern of invertebrate foraging is reported for white-faced capuchins in a dry forest in northwestern Costa Rica (Perry and Ordoñez-Jiménez 2006).

In conclusion, the feeding ecology of white-faced capuchins at EBLS appears to be most consistent with the reliance on highly selected foods rather than the use of fallback foods or keystone plant resources. This contention is supported by several lines of evidence. White-faced capuchins at EBLS depend on a small subset of plant resources each month (three species each month account on average for 64.6% of plant feeding time), and year-round (six tree species of five tree genera account for 52.7% of yearly feeding time). This is consistent with the contention of Sussman (1987). As suggested by Sussman (1987), in many primate taxa, a limited number of plant species with different morphological features account for most of the annual diet. Sussman (1987) referred to this as a species-specific dietary pattern. On the other hand, the exploitation of commonly used plant species of different eco-morphological characteristics does not necessarily require derived anatomical adaptations to increase foraging or feeding efficiency. For instance, this varies from soft fig berries to harder palm nuts. This differs from what was suggested for fallback foods in which for exploiting particular foods, specific dental and anatomical adaptations are required in order to access them (Marshall and Wrangham 2007).

The abundance of these trees is variable at EBLS. This suggests that the abundance of highly selected foods is site dependent, thus some trees may be relatively common or rare depending on the species. For example, at the EBLS *Bactris* was cultivated at the site in the 1980s for agricultural purposes, and therefore is a common tree species. It is a synchronously producing species and among the top three tree species fed across three months of this study. In contrast, based on a botanical survey, *Dipteryx* is a relatively rare tree species (0.5 ind/ha). Nonetheless, this species mast fruits and was consumed for seven months. Overall, the pattern exhibited by white-faced capuchins of exploiting a small set of tree species during the annual cycle enables these monkeys to efficiently locate and relocate feeding sites using spatial memory, and exploit different types of foods that pose different physical challenges.

At EBLS there was limited monthly variation in the behavioral and ranging patterns of the white-faced capuchin study group. White faced-capuchins at EBLS presented an ecological pattern based on the exploitation of a limited set of major feeding tree species (HSFs) which were available year-round (see above, Sussman 1987 and examples herein). They also devoted a relatively consistent amount of time per month to invertebrate prey (42.9% of their feeding and foraging time). Day range also changed little within and across months. Similarly, Stone (2004) found limited variation in squirrel monkeys (*Saimiri sciureus*) day range despite marked seasonal changes in diet. Squirrel monkeys switched from fruits to insects during the dry season, but retained a similar daily path length throughout the annual cycle. Garber (1993) also found that despite differences in food availability, saddleback tamarins (*Leontocebus fuscicollis*) and mustached tamarins (*S. mystax*) exhibited a consistent foraging and ranging pattern throughout the year. These included ~40% of monthly feeding time devoted to invertebrate foraging, a monthly mean day range of 2047 meters, an activity budget in which individuals were active approximately 10 hours per day, and consistency in the amount of time per month devoted to plant feeding. Consequently, Garber (1993) suggested that certain aspects of the feeding ecology, habitat use, and foraging patterns of wild primates remained highly stable throughout the annual cycle. He argued that the tamarins applied a similar set of foraging strategies or rules to exploit a range of different plant resources (ripe fruits, nectar, exudates). This appears to be similar to the foraging pattern exhibited by white-faced capuchins at EBLS. Given limited seasonality in temperature and rainfall at EBLS, the white-faced capuchins were characterized by low monthly variation in the time allocated to the exploitation of different food types. Although there were changes in the location and availability of food resources, white-faced capuchins applied a consistent set of foraging rules for the exploitation of different tree species and food types. Capuchins use spatiotemporal information on the distribution and availability of feeding trees to find and navigate to these sites. In general, white-faced capuchins at EBLS exhibited a similar behavioral repertoire as other *C. imitator* populations and other *Cebus* species. In sum, the foraging rules, ranging, and feeding ecology of the white-faced capuchins at EBLS remained relatively stable year-round.

Species	Feeding	Foraging	(Feed+Forage)	Resting	Social interact	Traveling	O. or U.	Study length	Location	References
Cebus cuscinus	-	-	(67)	11	-	22	-	2 months	Cocha-Cashu, Peru	Janson (1975, in Fresse and Oppenheimer 1981)
	22	39	-	18	-	21	-	12 months (1976–1977)	Cocha-Cashu, Peru	Terborgh (1983)
Average	22	39	-	14.5		21.5				
Cebus yuracus	10	54	(64)	5	6	25	-	12 months (2005–2006)	Tiputini Biological Station, Yasuní, Ecuador	Mathews (2008, 2009)[1]
Average	10	54	-	5	6	25	-			
Cebus imitator	-	-	(43.5)	26.1	1.58	11.6	0.3	15 months (1991, 1993–1996)	Refugio de Vida Silvestre Curú	Baker (1998)
	19.1	19.9	-	12.7	7.5	25.7	1.8	12 months (1993–1994)	Parque Nacional Santa Rosa, Costa Rica	Bergeson (1996)[2]
	13.4	-	-	-	9.4	16.2	-	15 months/ 836 hours (1980–1981)	Trujillo, Honduras	Buckley (1983)[3]
	-	-	(41)	21	1.5	37	-	9 months (1973)	Parque Nacional Santa Rosa, Costa Rica	Freese (Unpubl. in Fresse and Openheimer 1981)
	22	57	-	10	-	9	2	575 hours (1986–1987)	Barro Colorado Island, Panama	Mitchel (1989)[4]
	-	-	(50)	20	-	30	-	2 months (1982)	Refugio Silvestre R. L. Rodríguez Caballero, Palo Verde, Costa Rica	Moscow and Vaughan (1987)[5]
	-	-	(26.7)	16.1	12.9	45.9	-	18 months (1966–1967)	Barro Colorado Islan d, Panamá	Oppenheimer (1968)[6]
	37.9	15.1	-	16.2	8.6	19.5	2.3	1238 hours (1995–1996)	Parque Nacional Santa Rosa, Costa Rica	Rose (1998)[7]
	-	-	(55)	5	25	15	-	13 month (2009–2011)	Pacuare Nature Reserve, Costa Rica	Eadie (2012)
Average	23.1	30.7	-	15.9	9.5	23.3	2.0			
Cebus olivaceus	-	-	(38)	22	9	20	11	5 months (1980)	Hato Masaguaral, Venezuela	Fragaszy (1990)[8]
	19	17	(36)	23	3	38	-	485 hours, 11 months (1989–1991)	Hato Piñero, Venezuela	Miller (1992, 1996)[9]
	29.3	17.1	-	20.1	8.5	20.1	4.9	14 months (1977–1979)	Hato Masaguaral, Venezuela	Robinson (1986)
Average	24.2	17.1		21.7	6.8	26.0	8.0			
Cebus castaneus	-	-	24.1	-	-	74.9	-	3 months (1993	Station des Nouragues, French Guyana	Youlatos (1998)[10]
Average	-	-		-	-	74.5	-			
Cebus kaapori	24	16	(40)	9	3	48	-	6 months (2010)	Tucuruí Hydroelectric Dam, Pará, Brazil	Oliveira et al. (2014)
Average	24	16	-	9	3	48	-			
Average for genus Cebus	20.7	31.4		13.2	6.3	36.4	5.0			

Abbreviation: O. or U.: Others or Unknown

Notes: [1] His "foraging" category includes "insect, nut, and opportunistic fruit; [2]The data were combined from both sites; in addition, active posture is summed with foraging; [3]Social interactions are based on the sum of his "Groom" and "Play" categories only, therefore these data are approximated; [4]In this case, "social" and "rest" are pooled together by the author; [5]Approximate calculation from line graph; [6]Social interaction is the sum of his "play" and "allogrooming;" [7]Her "scan" and "forage" categories are pooled here as "foraging;" [8]In "other," I include their category "scan" that is defined as "visual inspection of the surrounding area without fixed gaze, turning head side to side. Scored while stationary or moving (Fragaszy 1990: 83). See also Fragaszy (1986); [9]Averages of the percentages of the females of the so-called "large group" and "small group" samples; the sum surpasses 100% greatly; [10]The calculations were performed with the number of bouts reported in his Table 1 for "travel" and feed/forage." The time allocated to travelling reported in this article was not used in the calculation because it appeared to be extremely large

Comments: (a) Because of the disparity in the presentation of the data by each author, only the major categories (except our "feed+forage") were used in the calculation of the averages. It is important to note that these averages must be used with caution because of differences between the studies in terms of field site characteristics, category definitions, study length, field methodology, and statistical analyses, among other aspects; (b) The data are presented in percentages; however, because of the disparities indicated previously, the sum may not reach exactly 100%; (c) The original table was published in the doctoral dissertation of Urbani (2009), it includes not only *Cebus* but also *Sapajus*; this is, -and was-, the first comprehensive review of the activity budget for both genera.

Table 4.21. Activity budget of wild untufted capuchin species at different field sites.

Table 4.22. Diet of wild untufted capuchin species at different field sites.

Species	Plant						Animals					Study length	Location	References
	Plant parts	(Fruit /Seed)	(Leaves)	(Shoots)	(Flowers)	(Others)	Animal matters	(Invertebrates)	(Vertebrates)	(Unknown)	Others			
Cebus albifrons	80	-	-	-	-	-	20	-	-	-	-	2 month (1977)	Territorio Faunístico El Tuparro, Colombia	Defler (1979)
Average	80	-	-	-	-	-	20	-	-	-	-			
Cebus cuscinus	100	(97)	-	(1)	-	(2)	-	-	-	-	-	12 months (1976- 1977)	Cocha-Cashu, Peru	Terborgh (1983)[1]
Average	100	-	-	-	-	-	-	-	-	-	-			
	56	(54)	-	-	(2)	-	41	(38)	(3)	-	5	15 months (1991, 1993–1996)	Refugio de Vida Silvestre Curú, Costa Rica	Baker (1998)
	43.4	-	-	-	-	-	55.7	(49.2)	(6.5)	-	-	12 months (1993- 1994)	Parque Nacional Santa Rosa and Reserva La Selva, Costa Rica	Bergeson (1996)[2]
	82.7	(81.2)	(1.3)	-	(0.2)	-	16.9	(16.9)	-	-	-	22 months (1983- 1986)	Parque Nacional Santa Rosa, Costa Rica	Chapman (1987)
	59.4	-	-	-	-	-	40.6	(40.6)	-	-	-	9 months (1973)	Parque Nacional Santa Rosa, Costa Rica	Freese (1977)[3]
	80	(65.9)	(7.6)	-	(4.1)	(2.4)	20	-	-	-	-	1966-1967	Barro Colorado Island, Panama	Hladik and Hladik (1969), Hladik et al. (1971)[4]
	73.8	(72.7)	-	-	-	(1.1)	23.3	(21.4)	(2)	-	3	12 months (1998)	Área de Conservación Guanacaste, Costa Rica	MacKinnon (2006)
	22	(20)	-	-	-	(2)	57	(57)	-	-	-	575 hours (1986-1987)	Barro Colorado Island, Panama	Mitchel (1989)[5]
Cebus imitator	80	-	-	-	-	-	20	(20)	-	-	-	2 months (1982)	Refugio Silvestre R. L. Rodríguez Caballero, Palo Verde, Costa Rica	Moscow and Vaughan (1987)
	80	(65)	-	-	-	(15)	20	(20)	-	-	-	18 months (1966–1967)	Barro Colorado Island, Panama	Oppenheimer (1968), Fresse and Openheimer (1981)
	45.5	-	-	-	-	-	48	47.4	0.5	-	-	11 months (1992–19939)	Lomas Barbudal, Costa Rica	Perry and Ordóñez-Jiménez (2007)
	63	(60)	(<1)	-	(2.5)	-	36	(35)	(1)	-	-	24 months/ 1238 hours (1995–1996)	Parque Nacional Santa Rosa, Costa Rica	Rose (1998)
	64	(57.2)	(1.4)	(2.5)	(2)	-	36.1	(21.4)	-	(114.7)	-	15 months/ 836 hours (1980–1981)	Trujillo, Honduras	Buckley (1983)[6]
	61.7	-	-	-	-	-	38.3	-	-	-	-	N/A-Dry season (Pre-1991)	Parque Nacional Santa Rosa, Costa Rica	Gilbert et al. (1991)
	62.5	-	-	-	-	-	37.5	-	-	-	-	13 month (2009–2011)	Pacuare Nature Reserve, Costa Rica	Eadie (2012)[7]

Species	Plant						Animals					Study length	Location	References
	Plant parts	(Fruit /Seed)	(Leaves)	(Shoots)	(Flowers)	(Others)	Animal matters	(Invertebrates)	(Vertebrates)	(Unknown)	Others			
Average	62.4	-	-	-	-	-	35.0	-	-	-	4			
Cebus olivaceus	91.6	(23.2)	(12.2)	(2.5)	(1.1)	(52.5)	8.4	(6.1)	(0.2)	(2.1)	-	5 months (1980)	Hato Masaguaral, Venezuela	Fragaszy (1986)[8]
	41.7	(41.7)	-	-	-	-	48.6	-	-	-	9.8	485 hours, 11 months (1989–1991)	Hato Piñero, Venezuela	Miller (1992)
	40.3	(36.7)	-	-	-	(3.8)	43.6	(43.6)	-	-	16.1	14 months (1977- 1979)	Hato Masaguaral, Venezuela	Robinson (1986)[9]
Average	57.9	-	-	-	-	-	33.5	-	-	-	13.0			
Cebus castaneus	66.7	(54.6)	(6.1)	-	(6.1)	-	33.3	(33.3)	-	-	-	12 months (1976- 1977)	Raleighvallen-Voltzberg Natural Reserve, Suriname	Mittermeier and van Roosmalen (1981)
Average	66.7	-	-	-	-	-	33.3	-	-	-	-			
Cebus kaapori	84	-	-	-	-	-	13	-	-	-	3	6 months (2010)	Tucuruí Hydroelectric Dam, Pará, Brazil	Oliveira *et al.* (2014)[10]
Average	84	-	-	-	-	-	13	-	-	-	3			
Average for genus Cebus	70.2	-	-	-	-	-	27.0	-	-	-	6.7			

Notes: [1]It was not possible to calculate vegetative items separately from the animal matter because the author does not discriminate beyond the category of feeding; [2]The data were combined from both sites, (P. N. Santa Rosa is a dry forest, and Reserva La Selva is a rainforest; [3]An average was estimated here for both seasons (dry and wet), and the percentage was also recalculated from the original numbers in Freese (1977); [4]The vegetative items are calculated from a bar graph in Hladik and Hladik (1969); [5]The remaining percentage is unknown where is allocated; [6]The category animal matter includes "unknown" animal matter and bouts of "Insects;" [7]Calculated from the percentage of her categories "foraging for fruits" and "foraging for insects;" [8]Her "woody" parts were included here as "Plant parts (Others);" [9]Calculated from a bar graph; [10]"Others" include "Flowers, leaves, branches, honeycomb, wasp nests, snails, a rodent and unidentified items" (Oliveira *et al.* 2014: 39).

Comments: (a) Because of the disparity of the presentation of the data by each author, only the two major categories (vegetative items and animal matters) were considered; the category "others" was not used in the calculation of the means. It is important to note that these averages must be used with caution because of differences between the studies in terms of field site characteristics; category definitions, study length, field methodology, and statistical analyses, among other aspects; (b) The data are presented in percentages; however, because of the disparities indicated previously, the sum may not reach exactly 100%; (c) The original table was published in the doctoral dissertation of Urbani (2009), it includes not only *Cebus* but also *Sapajus*; this is, -and was-, the first comprehensive review of this data for both genera.

Table 4.22. continued.

Species	Group size (range)	Number of groups	Home Range (ha)	Density (ind/km2)	Day range (m) (max-min)	Overlap range (%)	Location	References
Cebus albifrons	35	1	115 (110–120)	30.4*	4500 (4000–5000)	21.7	Parque Nacional El Tuparro, Colombia	Defler (1979, 1982)
Average	35	-	115	30.4	4500	21.7		
	10	10	-	9.3** (2–24)	-	-	Upper Nanay river, Samiria, Panguana, Cocha Cashu, Peru	Freese (1975), Neville *et al.* (1982)
Cebus cuscinus	13.5 (12–15)	-	65 (60–70)	37.5** (30–45) 20.8*	-	-	Cocha-Cashu, Peru	Janson (1975, in Freese and Oppenheimer 1981)
	15	1	≥150	>10**	1820±520	-	Cocha-Cashu, Peru	Terborgh (1983)
	10.3 (6–13)	-	-	11.3**	-	-	Lower Urubamba and Tambo rivers, Peru	Aquino et al. (2013)
	-	-	-	0.9**	-	-	Western Pando Department, Bolivia	Cameron et al. (1989)
Average	12.5	5.5	107.5	13.8	1820	-		
Cebus unicolor	25.4	-	>1200	16.5**	-	-	Urucu River, Amazonas, Brazil	Peres (1993)[1]
	-	-	-	5.1**	-	-	Jenaro Herrera region, Loreto, Peru	Aquino (1990)
Average	25.4	-	-	10.8	-	-		
	8	1	240	3.2*	-	-	Tiputini Biological Station, Yasuní, Ecuador	Mathews (2009)
	8.3 (5–10)	8	-	5.2** (4.2–6.2)	-	-	Cahuana Island, Reserva Nacional Pacaya-Samiria, Peru	Soini (1986)
Cebus yuracus	-	-	-	4.4**	-	-	Pucacuro river, Amazonia, Peru	Aquino *et al.* (1990)
	-	-	-	2.6** (1.8–3.6)	-	-	Quebradón el Ayo, Caño Pintadillo, and Caparúa Biological Station,,Caquetá River, Colombia	Palacios and Peres (2005)
Average	8.1	4.5	240	3.9	-	-		
	15 (7–37)	14	91	18.3** (16.7–20) 16.48*	1522.5 (400–2620; Wet: 1498, Dry:1547)	48.5	Refugio de Vida Silvestre Curú, Costa Rica	Baker (1998)
	28.5 (27–30)	1	36 (32–40)	79.2*	-	-	Hacienda Barqueta, Chiriquí, Panama	Baldwin and Baldwin (1976)
	14	1	102	2.8** 13.7*	-	0.02	Trujillo, Honduras	Buckley (1983)[2]
Cebus imitator	26	1	81.5 (78–85)	31.9*	-	-	Parque Nacional Santa Rosa, Costa Rica	Chapman (1988)
	15.7	19	150 (100–200)	10.5*	-	-	Parque Nacional Santa Rosa, Costa Rica	Fedigan *et al.* (1996)
	17	1	50	6.5** (5–7) 34*	-	-	Parque Nacional Santa Rosa, Costa Rica	Freese (1977), Freese and Oppenheimer (1981)
	14.6 (6–22)	2	~145 (60->300)	4.8**	2900 (2600–3200)	"occa-sional"	Parque Nacional Santa Rosa, Costa Rica	Rose (1998), Rose and MacKinnon in Pager (2002)

Table 4.23. Ranging patterns of wild untufted capuchin species at different field sites.

Species	Group size (range)	Number of groups	Home Range (ha)	Density (ind/km2)	Day range (m) (max-min)	Overlap range (%)	Location	References
Cebus imitator	33 (30–35)	3	358 (276–440)	9.2*	-	"approximately 10%"	Lomas Barbudal Biological Reserve, Costa Rica	Vogel (2005)
	20	4	90 (77–110)	22.2*	-	"overlaped so extensively"	Barro Colorado Island, Panama	Mitchel (1989)
	16	6	93	6.5*	-		Barro Colorado Island, Panama	Crofoot (2007)
	19	-	-	9.4**	-	-	Palo Verde, Costa Rica	Parger (2002)
	16	1	65	24.6*	4500±600		Refugio Silvestre R. L. Rodríguez Caballero, Palo Verde, Costa Rica	Moscow and Vaughan (1987)
	13 (11–15)	4	80	21** (18–24)	-	"to be minor"	Barro Colorado Island, Panama	Oppenheimer (1968), Freese and Oppenheimer (1981)
	21 (20–37)	1	128.6	3.7**	-	-	Lomas de Barbudal, Costa Rica	Perry in Parger (2002), Perry (1996)
	21	3	-	21.7**	-	-	Taboga Forest Reserve, Guanacaste, Costa Rica	Tinsley-Johnson et al. (2019)
Average	**19.3**	**4.4**	**113.1**	**17.9**	**2974**	**-**		
Cebus capucinus	6	1	-	13**	-	-	Arroyo Colosó, Sucre Department, Colombia	De la Ossa et al. (2013)
	3.6 (≤9)	-	-	13.5** (7.3–25.1)	-	-	Bosque de Yotoco Natural Reserve, Cauca, Colombia	Duque and Gómez-Posada (2009)
	-	-	-	170,6**	-	-	Parque Nacional Gorgona, Gorgona Island, Pacific Ocean, Colombia	Garcés-Restrepo et al. (2014)[3]
Average	**4.8**	**-**	**-**	**13.3**	**-**	**-**		
Cebus aequatorialis	9 (6–13)	-	-	3.7** (3.5–3.9)	-	-	8 sitesof the southc-central coastal region (Jama and El Palmar), Ecuador	Albuja and Arcos (2007)
	13.9 (5–20)	-	534 (507–561)	2.4** (2–22)	-	-	7 sites, northwestern Ecuador	Jack and Campos (2012)[4]
	5.2 (3–12)	-	-	0.3**	-	-	Noroeste Biosphere Reserve, Peru	Hurtado et al. (2016)
Average	**9.4**	**-**	**534**	**2.1**	**-**	**-**		
Cebus olivaceus	-	-	-	-	2083 (1824–2342)	-	Hato Piñero, Venezuela	Miller (1992)[5]
	≥10	20	257	3.9*	2141 (1046–3580)	-	Hato Masaguaral, Venezuela	Robinson (1986, 1988)
Average	**≥10**	**20**	**257**	**3.9**	**2112**	**-**		
Cebus versicolor	-	-	-	9.5** (3.8–15.3)	-	-	Cerro Bran (Puerto Rico), Departamento de Bolívar, Colombia	Green (1979)
	-	-	-	59.5** (2–117)	-	-	Inter-Andean forests, middle Magdalena River, Colombia	de Luna and Link (2018)
Average	**-**	**-**	**-**	**71**	**-**	**-**		

Table 4.23. continued.

Species	Group size (range)	Number of groups	Home Range (ha)	Density (ind/km2)	Day range (m) (max-min)	Overlap range (%)	Location	References
Cebus trinitatis	19 (8–30)	7	-	0.2** (0.07–0.23)	-	-	Trinity Hills and Bush-Bush sites, Trinidad Island	Agoramoorthy and Hsu (1995a, b)[6]
	10 (12–31)	6	>84	13.5**	-	-	Bush Bush, Trinity Hills, and Nariva Swamp, Trinidad Island	Phillips. and Abercrombie (2003)
Average	14.5	-	84	13.5	-	-		
Cebus castaneus	6 (1–20)	-	-	-	-	-	Amapá National Forest, State of Amapá, Brazil	Michalski et al. (2017)
Average	6	-	-	-	-	-		
Cebus kaapori	9	1	62	14.5*	2173 (400)	-	Tucuruí Hydroelectric Dam, Pará, Brazil	Oliveira et al. (2014)
	-	-	-	0.98**	-	-	Gurupi Biological Reserve, Maranhão, Brazil	Ferreira and Lopes (1996)
Average	9	1	62	14.9	2173	-		
Average for genus Cebus	14,4	3,9	179	19,2	2716	-		

Notes: [1]The home range reported in this article was not used in the calculation because it appeared to be extremely large; [2]The home range is from the main study group; [3]The density reported in this article was not used in the calculation because it appeared to be extremely large; [4]The home range reported in this article was not used in the calculation because it appeared to be extremely large; [5]The smallest day range is from the so-called "Small group" and the largest from the "Large group;" [6]The density was not taken into consideration in this table as the authors used hectares rather than km² in the calculation.

Comments: (a) Data used only when density or group size and home range are explicitly stated. References that only present information about group size (e. g. Southwick 1967) may be also checked in Oppenheimer and Freese (1981) and Fragaszy et al. (2004). The information from Thorington (1967) was not used because capuchins (Sapajus apella) were previously removed or poached in the area of his study; (b) *Density (ind./km²) calculated in this study using the number of individuals (group size) divided by the home range (in km²); (c) **Density (ind./km²) calculated by the original author. (c) Not used in the calculations because the value seems to be anomalous; (d) When the author presents more than one number for group size, day range, and density number, the values were summed and the average was calculated. (e) It is important to note that these averages must be used with caution because of differences between the studies in terms of field site characteristics, category definitions, study length, field methodology, and statistical analyses, among other aspects. The data are presented in percentages; however, because of the disparities indicated previously, the sum may not reach exactly 100%; (d) The original table was published in the doctoral dissertation of Urbani (2009), it includes not only Cebus but also Sapajus; this is, -and was-, the first comprehensive review of this data for both genera.

Table 4.23. continued.

5

Spatial Mapping in Wild White-Faced Capuchin Monkeys (*Cebus imitator*): A Natural Field Study

In this chapter, I examine patterns of ranging and spatial memory in a group of white-faced capuchin monkeys (*Cebus imitator*). A primary objective of this chapter is to determine the degree to which wild capuchins use a coordinate-based or route-based spatial representation to locate resting and feeding trees.

Poucet (1993) proposed a model that can be used to empirically test the ability of foragers to form different spatial representations in larger and smaller scale space. Poucet argued that in larger scale space, foragers may use a route-based spatial representation in which the re-use of travel routes and landmarks or other topographical features of the habitat are encoded as "places" for navigation. In smaller scale space, however, foragers are expected to encode feeding and resting sites in a coordinate-based spatial representation in which each location is remembered as a set of coordinates which represent information of exact distances and directions. Using coordinated information and different views of the same goal, the forager is able to compute new shortcuts to reach their goals.

In this chapter, the following questions are explored: (a) how many feeding/resting sites do capuchins visit each day?, (b) what is the distance between sequential feeding/resting sites?, (c) how frequently do capuchins use and re-use the same route segments, (d) do capuchins use landmarks (nodes) for re-orienting travel?, and (e) is capuchins navigation within large-scale space more consistent with a coordinate-based or a route-based spatial representation?.

Results

Pattern of feeding/resting tree visits and distribution

A question in the study of spatial mapping focuses on the pattern in which feeding/resting sites are visited. Thus, I examine how many new trees capuchins visit per day and per month. Capuchins fed and rested in an average of 7.8±0.7 individual trees per day (feeding trees: 6.6±0.6, resting trees: 1.7±0.7; Table 5.1). However, given that 2.4±0.6 of these trees were revisited on the same day, the capuchins fed or rested in a total of 10.2±0.8 trees per day. Urbani (2009: Appendix E) compiles monthly tables of trees used by white-faced capuchins and presents details on individual tree visits and tree IDs. Table 5.1 summarizes these results. On average the capuchins visited 31.8±8.5 different trees per month, with 37.5% of these trees comprising only three species (Table 5.2). The largest number of total trees were visited in May (n=45) and the smallest number was visited in December (n=15).

In order to examine the manner in which capuchins incorporate new trees into their foraging routes, I analyzed data on the discovery of novel feeding/resting sites. I found that on average, 27.5% of the trees visited each month had not been previously used. Monthly differences in the proportion of new trees used were not significant, ranging from 22.4% in June to 38.1% in March (Table 5.3; χ^2=6.68, d.f.=7, p>0.05). The data also suggest that the capuchins take advantage of the fruiting pattern of certain tree species and rely on a search pattern in which more than one tree of that tree species was visited on the same day (mean=1.6±0.5). This occurred on 62 days of the 94 observation days.

To determine the distribution of feeding/resting sites, the capuchin home range (23.7 ha on the EBLS property) was divided into 553 quadrats (20 × 20 m). One-fourth of these quadrats contained at least one white-faced capuchin feeding/resting tree (135/553). Of these 135 quadrats, 78.5% (106/135) contained one feeding/resting tree, 17.8% (24/135) contained two feeding/resting trees, and only 2.7% (5/135) contained three feeding/resting trees.

A major feeding/resting tree was defined as a tree in which the monkeys spent a minimum of 1% of their total feeding and resting time during the study period (after Jelinek and Garber; see Chapter 3). Overall, only 1.6% of quadrats (9/553) contained a major feeding or resting tree. The location of the major feeding/resting trees within the white-faced capuchin home range is illustrated in Fig. 5.1. Table 5.4 presents data on the use of major feeding/resting trees per month. The total area bounded by these major feeding/resting trees measured 5.6 ha (23.6% of the total of 23.7 ha of the "small forest"). This area was calculated using a closed polygon and represents approximately one-fourth of the size of the forest. However, given that these trees occupied only 9 quadrats, the area they occupied (0.4 ha) represents <5% of the group home range. I also calculated the area covered by the canopy of the major feeding/resting trees used in the "small forest." This was accomplished by measuring the diameter of the tree crown in two cardinal directions (3726 m²; Table 5.5). Major feeding/resting trees represented 1.7% of the total area of this forest (0.4 ha/23.7 ha; Table 5.5). These data suggest that white-faced capuchins are highly selective in their choice of major resting and feeding sites and that these trees are patchily distributed in the forest.

	Day number	1	2	3	4	5	6	7	8	9	10	11	12	Average	SD (±)	Min	Max
March	Resting tree visits	1	1	0	2	0	4	1	2	2	1	0	1	1.3	1.1	0	4
	Feeding tree visits	6	8	6	8	7	11	10	9	7	6	8	9	7.9	1.6	6	11
	Individual tree visits	7	9	6	10	7	14	11	11	8	6	8	10	8.9	2.4	6	11
	Total visits/day	13	17	9	17	9	15	13	15	9	9	12	10	12.3	3.1	9	15
	Revisits/day	6	8	3	7	2	1	2	4	1	3	4	0	3.4	2.5	0	8
April	Resting tree visits	4	4	0	0	0	1	1	1	1	2	0	1	1.3	1.4	0	4
	Feeding tree visits	6	6	5	9	5	10	9	4	7	7	4	6	6.5	2.0	4	10
	Individual tree visits	10	10	5	9	5	11	10	5	8	9	4	7	7.8	2.5	4	11
	Total visits/day	13	17	8	12	8	13	14	9	14	15	7	8	11.5	3.3	7	17
	Revisits/day	3	7	3	3	3	2	4	4	6	6	3	1	3.8	1.8	2	7
May	Resting tree visits	2	3	4	2	3	3	2	2	3	2	1	0	2.3	1	0	4
	Feeding tree visits	9	12	13	10	10	10	8	13	10	6	9	7	9.8	2	6	13
	Individual tree visits	11	15	17	12	13	13	10	15	13	8	10	7	12.0	3	7	17
	Total visits/day	12	17	20	14	17	17	13	18	14	13	10	7	14.3	4	7	20
	Revisits/day	1	2	3	2	4	4	3	3	1	5	0	0	2.3	2	0	5
June	Resting tree visits	2	3	4	4	2	2	1	5	1	3	3	2	2.7	1	1	4
	Feeding tree visits	9	10	12	11	9	12	11	13	10	7	9	8	10.1	2	9	13
	Individual tree visits	9	10	15	11	9	13	11	14	11	8	9	8	10.7	2	8	15
	Total visits/day	11	12	17	18	12	16	13	17	13	11	12	10	13.5	3	10	18
	Revisits/day	2	2	2	7	3	3	2	3	2	3	3	2	2.8	1	2	3
July	Resting tree visits	1	0	1	0	1	2	2	2	0	3	1	1	1.2	1	0	3
	Feeding tree visits	2	5	8	6	2	6	10	3	7	5	5	4	5.3	2	2	10
	Individual tree visits	3	5	9	6	3	8	12	5	7	8	6	5	6.4	3	3	12
	Total visits/day	6	6	9	6	7	8	12	5	8	8	6	9	7.5	2	6	12
	Revisits/day	3	1	0	0	4	0	0	0	1	0	0	4	1.1	2	0	4
August	Resting tree visits	0	0	1	1	1	2	4	2	2	2	2	1	1.5	1	0	4
	Feeding tree visits	4	11	9	10	10	9	5	3	3	2	4	3	6.1	3	2	11
	Individual tree visits	4	11	10	11	10	10	9	4	4	3	5	3	7.0	3	3	11
	Total visits/day	4	14	12	15	14	15	13	8	10	10	8	5	10.7	4	4	15
	Revisits/day	0	3	2	4	4	5	4	4	6	7	3	2	3.7	2	0	7
October	Resting tree visits	5	5	1	0	2	3	1	2	2	2			2.3	2	0	5
	Feeding tree visits	5	6	5	3	6	4	2	6	8	7			5.2	2	2	8
	Individual tree visits	8	9	6	3	6	6	3	8	9	8			6.6	2	3	9
	Total visits/day	10	10	7	4	8	6	4	12	10	9			8.0	3	4	12
	Revisits/day	2	1	1	1	2	0	1	4	1	1			1.4	1	1	4
December	Resting tree visits	1	0	2	1	0	0	2	0	4	0	0	2	1.0	1	0	4
	Feeding tree visits	2	3	2	1	4	1	1	4	3	3	1	2	2.3	1	1	4
	Individual tree visits	3	3	3	2	4	1	3	4	6	3	1	4	3.1	1	2	6
	Total visits/day	3	3	4	3	5	1	3	5	6	3	1	5	3.5	2	1	6
	Revisits/day	0	0	1	1	1	0	0	1	0	0	0	1	0.4	1	0	1
Total	Resting tree visits													1.7	0.4	0	4
	Feeding tree visits													6.6	0.6	1	13
	Individual tree visits													7.8	0.7	2	17
	Total visits/day													10.2	0.8	10	20
	Revisits/day													2.4	0.6	0	8

Note: The dataset of this table includes feeding/resting trees located within and outside the "small forest" of the EBLS

Table 5.1. Patterns of tree visits of the La Yunai group during the natural field study.

	Number of feeding trees used during the month	Number of feeding trees of the top 3 tree species used during each month	% of number of feeding trees of the top three tree species used during the month	1st top tree species	2nd top tree species	3rd top tree species
March	37	12	32.4	Dipteryx panamensis (88.9%; 8/9)	Inga spectabilis (33.3%; 2/6)	Sterculia recordiana (75%; 2/3)
April	30	11	36.7	Dipteryx panamensis (44.4%; 4/9)	Inga spectabilis (66.7; 4/6)	Ficus americana (20%; 3/15)
May	45	15	33.3	Miconia affinis (66.7; 10/15)	Ficus americana (20%; 3/15)	Dipteryx panamensis (22.2%; 2/9)
June	42	8	19.0	Ficus americana (20%; 3/15)	Brosimum alicastrum (60.7%; 3/5)	Ficus pertusa (50%; 2/4)
July	35	13	37.1	Inga spectabilis (33.3%; 2/6)	Miconia affinis (46.7%; 7/15)	Piper sancti-felici (100%; 4/4)
August	40	21	52.5	Ficus insipida (83.3%; 5/6)	Psidium guajava (75%; 9/12)	Bactris gasipaes (31.8%; 7/22)
October	34	12	35.3	Nephelium lappace-um (66.7; 2/3)	Bactris gasipaes (40.9%; 9/22)	Dipteryx panamensis (11.1%; 1/9)
December	13	7	53.8	Alchornea costari-censis (25%; 2/8)	Inga ruziana (66.7%; 2/3)	Cocos nucifera (100%; 3/3)
Average	34.5	12.4	37.5	-	-	-
SD (±)	9.9	4.34	11.2	-	-	-
Min.	13	7	19	-	-	-
Max.	45	21	53.8	-	-	-

Note: See Table 4.12. In parentheses, number of used feeding trees/Total number of individual feeding trees of the same given species, and proportion (percentages) of the used feeding trees vs. the total number of individual feeding trees of the same given species.

Table 5.2. Top tree species used per month.

	Day number	1	2	3	4	5	6	7	8	9	10	11	12	Average	SD (±)	Min	Max
March	New visited tree	7	8	5	4	1	4	3	4	2	2	1	3	3.7	2,2	1	8
	% New visited tree/ Ind. tree visits	-	89	83	40	14	29	27	36	25	33	13	30	38.1	25	13	89
April	New visited tree	10	4	1	4	0	3	3	1	1	2	1	1	2.6	2,7	1	10
	% New visited tree/ Ind. tree visits	-	40	20	44	0	27	30	20	13	22	25	14	23.3	12	0	44
May	New visited tree	11	10	3	4	4	1	2	4	3	0	3	1	3.8	3	0	11
	% New visited tree/ Ind. tree visits	-	67	18	33	31	8	20	27	23	0	30	14	24.6	17	0	67
June	New visited tree	9	5	4	1	1	3	2	4	2	2	1	2	3.0	2	1	9
	% New visited tree/ Ind. tree visits	-	50	27	9	11	23	18	29	18	25	11	25	22.4	11	9	50
July	New visited tree	3	2	1	3	0	3	4	1	2	3	0	3	2.1	1	0	4
	% New visited tree/ Ind. tree visits	-	40	11	50	0	38	33	20	29	38	0	60	28.9	19	0	60
August	New visited tree	4	4	7	4	2	1	2	2	0	1	2	0	2.4	2	0	7
	% New visited tree/ Ind. tree visits	-	36	70	36	20	10	22	50	0	33	40	0	28.9	21	0	70
October	New visited tree	8	3	0	2	3	2	0	2	3	2	-	-	2.5	2	0	8
	% New visited tree/ Ind. tree visits	-	33	0	67	50	33	0	25	33	25	-	-	29.6	21	0	67
December	New visited tree	3	0	2	1	0	0	1	2	1	3	0	2	1.3	1	0	3
	% New visited tree/ Ind. tree visits	-	0	67	50	0	0	33	50	17	0	0	50	24.2	26	0	67
Total	New visited tree													2.7	0.7	0	11
	% New visited tree/ Ind. tree visits													27.5	5.5	0	89

Table 5.3. New visited trees used by members of the La Yunai group during the natural field study.

Tree #	March	April	May	June	July	August	October	December
T1	X	X	X	X	-	-	X	-
T9	X	X	X	-	-	-	-	-
T10	X	X	X	X	X	-	X	-
T12	X	-	-	-	-	-	-	-
T32	X	-	X	X	-	-	-	-
T77	-	-	X	X	-	-	-	-
T94	-	-	X	-	-	-	-	-
T114	-	-	-	X	-	X	X	-
T182	-	-	-	-	-	X	-	-
Total	5	3	6	5	1	2	3	0

Table 5.4. Use of major feeding/resting trees per month.

Tree #	Focal activity	% total feeding/ resting frequency of use	Tree species	Crown radius (m)	Crown area (m2)	Crown area (ha)	% of area (ha)
T1	Feeding	3.8	Dipteryx panamensis	18.0	1017	0.10	0.43
T9	Feeding	6.4	Dipteryx panamensis	14.5	660	0.07	0.28
T10	Resting	3.0	Stryphnodendron microstachyum	13.0	531	0.05	0.22
T12	Feeding	1.6	Sterculia recordiana	6.0	113	0.01	0.05
T32	Feeding/ Resting	1.0	Brosimum alicastrum	14.0	615	0.06	0.26
T77	Feeding	1.3	Miconia affinis	4.5	64	0.01	0.03
T94	Feeding	1.5	Ficus americana	7.0	154	0.02	0.06
T114	Feeding/ Resting	1.9	Nephelium lappaceum	9.0	254	0.03	0.11
T182	Feeding	1.7	Ficus insipida	10.0	314	0.03	0.13
Average				10.7	414	0.04	0.17
SD(±)				4.5	315	0.03	0.13
Min				4.5	64	0.01	0.03
Max				18.0	1017	0.10	0.43

Note: the percentage is in relation to the total area of the "small forest" (23.74 ha).

Table 5.5. Area of the major feeding/resting trees used by members of the La Yunai group during the natural field study.

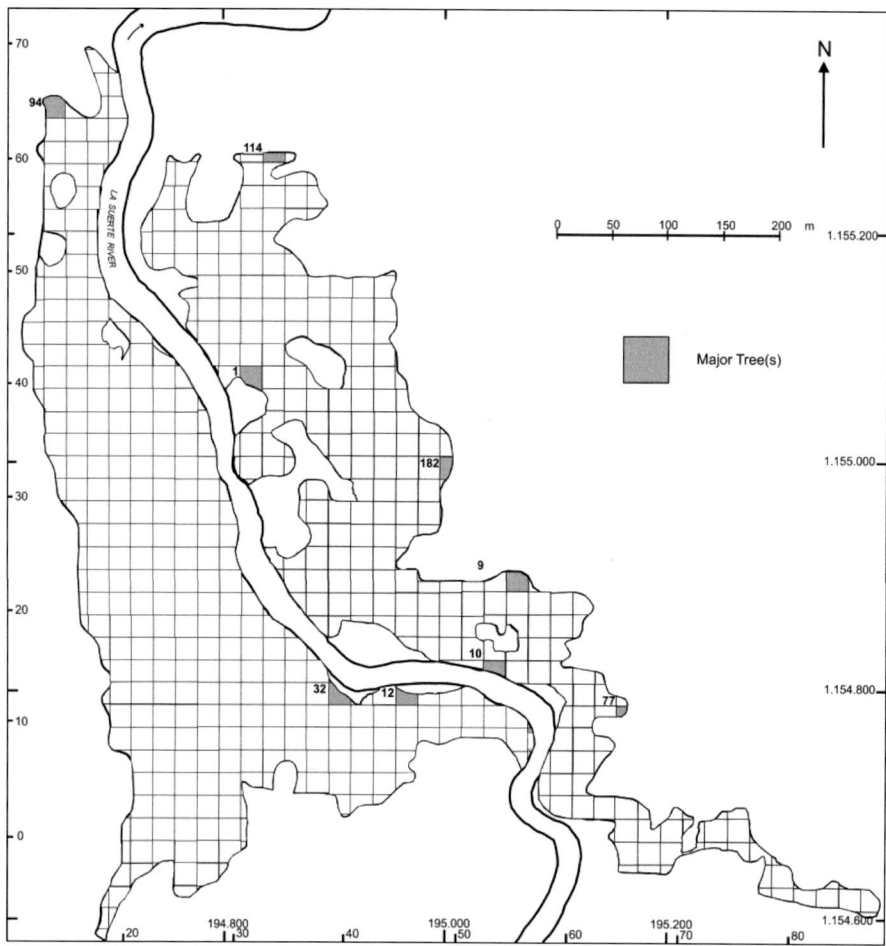

Figure 5.1. Major feeding/resting trees used by the La Yunai group during the natural field study.

Canopy density and forest profile: Field-of-view in the forest canopy

In order to test spatial memory in this capuchin group, I examined the degree to which the monkeys could sight directly between targets. Field-of-view was defined as an estimate of the monkey's unobstructed view across a distance of 20 m in the high, middle, or upper canopy levels (Garber and Jelinek 2006; see Chapter 3). In habitats in which primates have a large field of view such as savannas, a forager is expected to have unobstructed views at distances of hundreds of meters. In habitats in which primates have a limited field-of-view such as rainforests, a forager is expected to have a reduced field-of-view due to dense canopy. This measurement was used to determine the degree to which white-faced capuchin monkeys in a feeding/resting tree could sight directly to the next feeding or resting tree. In the case of white-faced capuchins, I calculated the field-of-view or the degree of unobstructed views at different levels within the canopy at a distance of 20 m from a major feeding/resting tree.

Tree #	Focal activity	DBH (cm)	Crown diameter (m)	Crown height (m)	Tree height (m)	Crown volume (m³)
T1	Feeding	100	36	20	38	3020
T9	Feeding	171	29	26	42	2250
T10	Resting	55	26	16	30	1230
T12	Feeding	55	12	16	20	88
T32	Feeding/ Resting	215	28	18	40	3150
T77	Feeding	12	9	7	14	73
T94	Feeding	66	14	10	22	440
T114	Feeding/ Resting	51	18	4	23	1230
T182	Feeding	98	20	22	28	361
Average		91	21	15	29	1316
SD (±)		64	9	7	10	1219
Min		12	9	4	14	73
Max		215	36	26	42	3150

Note: Major used trees are defined as the ones where the capuchins spent at least 1% of the total feeding and foraging time during the months of the natural field study. For this table, only major trees used within the EBLS property were considered.

Table 5.6. Description of the major individual feeding/resting trees utilized by La Yunai group during the natural field study.

In Table 5.6, I provide detailed information on tree and crown heights, crown diameter, and the crown volume of each of the major feeding/resting trees. The 9 most commonly used feeding/resting trees used by the white-faced capuchins had a mean tree height of 29±10 m, a mean crown diameter of 21±9 m, a mean DBH of 91±64 cm, and a mean crown volume of 1316±1219 m³ (Table 5.6). In Table 5.7, I also provide data on DBH, crown diameter, crown height, and tree height of the trees located within a 20 m radius of these most commonly used trees. The results indicate that the major feeding/resting trees were taller in height (29±10 m $vs.$ 16.2±4.4; z=3.83, d.f.=375, p<0.01), had larger crown diameter (21±9 m $vs.$ 6,8±3.3 m; z=4.73, d.f.=375, p<0.01), and greater DBH (91±64 m $vs.$ 28.1±22.8 m; z=2.94, d.f.=375, p<0.01) than the trees surrounding them. On average there were 40.9±19.1 trees located within a radius of 20 m from a major feeding/resting tree. The number of trees per radius varied from 16 to 73 trees (Table 5.7). Overall, 21.8% of trees within this sample plots ranged from 20 m and ≥25 m height.

I calculated the field-of-view or degree of visibility from major feeding/resting trees. This was accomplished by measuring the DBH, crown diameter, crown height, and tree height of all trees located within a radius of 20 m from the target tree. For each canopy level, I summed the number of trees and crown area that reached this height. The percentage of visual obstruction is the measure of the volume of tree crowns at each level of the canopy that would block a forager's field of vision at that height. These results indicate that on average at a height of 0–14 m, 84.2% of the field of view was obstructed for white-faced capuchins. At a height of 15–19 m, 52.4% of the white-faced capuchin field-of-view was obstructed. At a height of 20–24 m, the degree of obstruction was 22.0%, and

Plot of tree #	# of trees		DBH (cm)	Crown Diameter (m)	Crown Height (m)	Tree Height (m)
T1	47	Average	26.9	6.9	12.3	18.2
		SD (±)	14.0	3.1	5.1	5.8
		Max	75	20	25	33
		Min	10	3	3	11
T9	61	Average	20.9	5.8	10.4	14.5
		SD (±)	13.2	2.2	2.6	3.1
		Max	92	13	17	22
		Min	10	2	3	9
T10	41	Average	34.3	9.0	11.1	19.3
		SD (±)	22.4	4.1	4.1	4.4
		Max	111	20	20	29
		Min	10	3	2	11
T12	27	Average	30.76	6.8	13.5	17.0
		SD (±)	18.49	3.7	7.2	3.8
		Max	85	20	44	27
		Min	10	1.5	6	11
T32	33	Average	32.2	8.1	10.8	19.2
		SD (±)	20.3	4.4	6.0	6.6
		Max	89	20	24	35
		Min	11	1	1	6
T77	51	Average	19.3	5.9	8.0	13.8
		SD (±)	9.3	2.4	3.1	3.1
		Max	56	12	14	19
		Min	10	1	2	8
T94	19	Average	25.1	7.5	9.3	16.1
		SD (±)	23.1	4.5	3.7	5.2
		Max	105	21	16	32
		Min	11	2	3	9
T114	16	Average	27.8	4.4	4.25	10.6
		SD (±)	13.8	1.0	2.3	2.5
		Max	63	6	8	15
		Min	13	3	1	5
T182	73	Average	35.8	6.8	9.7	16.0
		SD (±)	70.2	4.0	4.3	4.9
		Max	583	26	30	42
		Min	10	2	1	8
Total	368	Average	28.1	6.8	10.2	16.2
		SD (±)	22.8	3.3	4.3	4.4
		Max	583	26	44	42
		Min	10	1	1	5

Table 5.7. Description of the plot around the major feeding/resting trees used by the La Yunai group during the natural field study.

| NAVIGATING WITH WHITE-FACED CAPUCHIN MONKEYS

Plot of tree #	# of trees	% 0-<15 m	%15-<20 m	% 20-<25 m	% ≥25 m
T1	47	29.8	38.3	14.9	17.0
T9	61	50.8	41.0	8.2	0
T10	41	12.2	43.9	31.7	12.2
T12	27	37.0	37.0	18.5	7.4
T32	33	21.2	30.3	30.3	18.2
T77	51	58.8	41.2	0	0
T94	19	42.1	31.6	21.1	5.3
T114	16	93.8	6.3	0	0
T182	73	41.1	48.0	5.5	5.5
Average	41	43.0	35.3	14.5	7.3
SD (±)	19	23.8	12.2	12.0	7.1
Max	73	93.8	48	31.7	18.2
Min	16	12.2	6.3	0	0

Table 5.8. Tree height categories and tree number of plots around major feeding/resting trees used by the La Yunai group during the natural field study.

when positioned at a height of ≥25 m, 16.5% of capuchins' visual field was obstructed by the surrounding vegetation (Table 5.8). These data suggest that when feeding/foraging in the uppermost parts of the tree crown, capuchins had a relatively unobstructed view for at least 20 m. However, when feeding/foraging at a height lower than 20 m, their field-of-view was more limited. These data also suggest that when capuchins travel in the understory and the lower and mid-canopy levels to feeding/resting that are ≥20 m apart, they cannot see directly to the next tree and must rely on landmark cues or a coordinate internal spatial map to efficiently encounter previously visited feeding/resting sites. Although I did not collect data on capuchins' travel height, studies of *C. imitator* and other capuchins species indicate that these primates traveling and foraging occur mainly in the lower portion of the canopy and in the understory (Mittermeier and van Roosmalen 1981, Gebo 1992, Bergeson 1996, Youlatos 1998, Urbani, pers. obs.).

A test of spatial memory: The case of the use of two feeding tree species

To examine if capuchins select the nearest neighbor trees of a target tree species as the next feeding site, I collected data on the sequential use of feeding sites, focusing on two tree species, *Dipteryx panamensis* and *Sterculia recordiana*. *D. panamensis* is exploited by white-faced capuchins year-round and *S. recordiana* is exploited only during the month of March (Tables 5.9, 5.10). All fruiting adult trees of these two tree species in the home range of the capuchin study group were identified, and their location was plotted on the field map. There were 11 individual trees of *Dipterix* and six individual trees of *Sterculia*. These tree species were selected because they represent large and highly productive feeding sites (*Dipteryx* accounted for a total of 14% of capuchin total plant feeding time, and *Sterculia* accounted for 2.2% of total plant feeding time). *Dipteryx* was the focus of capuchin ranging behavior across seven months of the 8-month study period and visited on 45 of 94 observation days. *D. panamensis* was also consumed as one of the top three feeding species in this study period. The number of *Dipteryx* trees visited per day was 1.8±1.2 and the number of revisits during the same day was 1.1±0.3. In the case of *Sterculia*, individual trees were visited on nine days. The number of trees visited per day was 1.1±0.3 and the number of revisits on the same day was 1. Given that capuchins visited different *Dipteryx* and *Sterculia* trees on the same day and these trees were visited in succession, I examined patterns of tree selection and inter-tree distance in order to test the hypothesis that capuchins encoded the location of these 17 trees, and visited them using a distance-minimizing principle.

Beginning with the location of a previous feeding/resting tree, I found that the capuchins visited the nearest tree of *Dipteryx* or *Sterculia* in 85.9% of cases (55/64). In the remaining 14.1% (9/64) of cases, the second nearest tree of *Dipteryx* or *Sterculia* was selected. The mean straight-line distance to the nearest neighbor *Dipteryx* tree was 70±78m. The mean straight-line distance to the nearest neighbor *Sterculia* tree was 121±70m. When traveling to the nearest *Dipteryx* tree, the mean distance traveled by the capuchins was 91.3±117 m. An index of circuity was calculated to compare the distance white-faced capuchins traveled between sequential feeding/foraging sites and the straight-line distance between these sites. A circuity index (CI) of 1 indicates that travel is straight-line or direct. A circuity index of 1.1 means that the forager traveled 10% above the most direct route between two sites. A CI <1.10 often is considered consistent with direct travel due to the fact that distance deviation is minimal (Garber and Jelinek 2006, Normand and Boesch 2009). The circuity index was 1.26.

The mean distance traveled by white-faced capuchins to reach the nearest neighbor *Sterculia* tree was 182±118 m. The circuity index was 1.43. When the capuchins did not choose the nearest neighbor tree, they traveled 138±87 m (*Dipteryx*) and 326±306 m (*Sterculia*). The straight-line distance between a given tree and the second nearest tree was 104±60 m (*Dipteryx*; CI=1.32) and 188±161 m (*Sterculia*; CI=1.64). The ability of the capuchins to locate the nearest neighbor *Dipteryx* and *Sterculia* trees resulted in a decrease of travel distance of 33.8% (138m/91.3m) and 44.2% (326 m/182 m) respectively. The choice of the nearest and second nearest neighbor tree implies that capuchins do not rely on a random search. These results also indicate that capuchins exploit important feeding trees such as *Dipteryx* and *Sterculia* by recalling information concerning the location of many trees in their home ranges and forming a foraging rule in which the nearest trees are selected first.

Evaluating travel itineraries

A relevant question in the study of capuchin spatial mapping is the degree to which monkeys travel using a distance-limiting pattern to reach sequential feeding/resting sites and foraging sites. To examine this, I measured the distances traveled between sequential foraging sites and feeding/resting trees throughout the day. The mean straight-line distance between those sites was 111±81 m. Overall, the distance capuchins traveled between these sequential sites was 158±122m (Table 5.11). The mean circuity index was 1.42 (*n*=335). In 22.7% of cases, CI was <1.10 (76/335). In 26.6% of cases, CI ranged between 1.10 and ≤1.25 (90/335). In 23.9% of cases, CI was >1.25–1.50 (81/335), and in the remaining 26.9% of

Dipteryx panamensis	T1	T2	T9	T13	T15	T22	T30	T33	T160	AA-Dip	BB-Dip
T1 to	0	142	298	275	341	357	343	58√√	308	119	205
T2 to	142√	0	387	409	481	498	476	183	396	257	187
T9 to	298	387	0	187	271	294	213	240	9	302	252
T13 to	275	409	187	0	92X	115	69	226	191	204	355
T15 to	341	481	271	92X	0	24√	68	301	274	247	444
T22 to	357	498	294	115	24	0	90	319	297	258	466
T30 to	343	476	213	69	68√√	90	0	294	214	268	413
T33 to	58	183	240	226	301	319	294	0	250	116	187
T160 to	308	396	9	191	274	297	214	250	0	311	259
AA-Dip to	119	257	302	204	247	258	268	116	311	0	303
BB-Dip to	205	187	252	355	444	466	413	187	259	303	0

Sterculia recordiana	T4	T7	T12	~~AA-Ste~~	BB-Ste	CC-Ste
T4 to	0	55	345	~~51~~	281	303
T7 to	55√√√√	0	399	~~28~~	329	354
T12 to	345	399	0	~~394~~	121	68
~~AA-Ste to~~	~~51~~	~~28~~	~~394~~	0	~~332~~	~~353~~
BB-Ste to	281	329	121	~~332~~	0	53
CC-Ste to	303	354	68	~~353~~	53	0

Abbreviations: Trees indicated with a "T" are feeding trees (e.g. T22); trees indicated with a double letter AA, BB, or CC are trees not used by the capuchins (e.g. BB-Ste); *Dipteryx panamensis* (Dip); *Sterculia recordiana* (Ste); Shorter distance used (√); Longer distance used (X). Each √ symbol used implies a single visit.
Notes: 1) Numbers within the table are the distances between both individual trees (m); 2) Trees with dashed line were not fruiting at the time of the study (e.g. AA-Ste); 3) Trees in the first column are the starting trees and the trees across the row are the arriving trees; 4) *n*=12 cases.

Table 5.9. Inter-distance between trees (*Dipteryx panamensis* and *Sterculia recordiana*) utilized by the La Yunai group during the natural field study.

	# (%)	of used	tree	individuals					Total # of tree
	March	April	May	June	July	August	October		individuals
Dipteryx panamensis	8 (72.7%)	4 (36.4%)	5 (45.5%)	1 (9.1%)	2 (18.2%)	1 (9.1%)	1 (9.1%)		11
Sterculia recordiana	3 (50%)	-	-	-	-	-	-		6

Table 5.10. Number of individual trees of *Dipteryx* and *Sterculia* trees used per month.

cases, CI was ≥1.50 (91/335). Considering the equal distribution in CI classes suggest that minimizing distance is not significant at this distance.

To explore how insect feeding/foraging affected white-faced capuchin travel routes, I calculate the mean circuity index when capuchins traveled between feeding/resting trees, between foraging patches (see definition in Table 5.11), from foraging patches to feeding/resting trees, and from feeding/resting trees to foraging patches. This was done in order to determine if goal-directed travel was associated with insect procurement. Overall, the capuchins traveled 41% more than the straight-line distance between feeding/

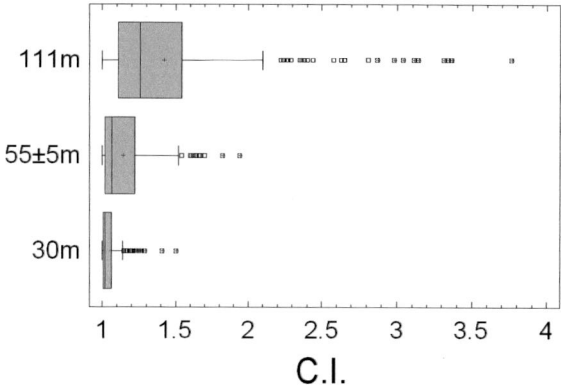

Figure 5.2. Comparison of the index of circuity at different distances.

resting trees, 53% more than the straight-line distance between insect foraging patches, 54% more than the straight-line distance from foraging patches to feeding/resting trees, and 37% more than the straight-line distance from feeding/resting trees to insect foraging patches (Table 5.11).

To test if capuchins adjust the circuity index while progressing toward their goal, I also calculated the index of circuity at half (55±5 m) the distance and at 30 m to the target. At 55±5 m, the mean CI was 1.14. Given the denseness of the forest canopy, at this distance, it was not possible for white-faced capuchins to see their targets or goals. In 57.1% of these cases, CI was <1.10 (149/261). In 21.8% of cases, CI was between 1.10 and ≤1.25 (57/261).

I also calculated the circuity index when capuchins traveled the final 30 m to reach the perimeter of feeding/resting trees. This was accomplished as follows. The mean crown radius of feeding/resting trees in the capuchins´ home range was 5.9 m. The average crown diameter of the trees around the major feeding/resting trees in the capuchins´ home range was 6.8 m. Thus, I assume that within a distance of 12.7 m (the sum of 5.9 m and 6.8 m) the capuchins were sufficiently near to their target, that they could see the tree directly. I also assumed that at twice this distance ≥25 m they could not sight directly to their goal and would be forced to rely on spatial information to reach the target tree.

At a distance of 30 m, the capuchins were highly efficient in selecting direct routes. From the total of 120 recorded cases, the average circuity index was 1.06. In 79.2% of the cases, CI was <1.10 (95/120). In 16.7% of cases, CI ranged between 1.10 and ≤1.25 (20/120), and in the remaining 4.2% of the cases, CI was >1.25 (5/120). These results suggest that once in the immediate vicinity of a feeding/resting tree, capuchins take highly direct routes of travel.

To study the relationship between CI and the distance traveled to reach a target, I compare CI at different distances. When traveling 111 m, CI was 1.42. When traveling 55±5 m, CI was 1.14. When traveling 30 m, CI was 1.06. Differences between CIs at distances of 111 m and 55±5 m were significant, and indicate that at greater distances capuchins were less efficient in traveling to their goals ($z=10.12$, d.f.=594, $p<0.01$). When comparing the CI at 55±5 m with the circuity index at 30 m, differences also were significant ($z=6.08$, d.f.=383, $p<0.01$). These results are consistent with the idea that capuchins efficiently navigate near targets, or encoded a coordinated-based spatial representation to reach feeding/resting sites. This pattern is shown in Fig. 5.2.

Table 5.11. Indices of circuity of the La Yunai group during the natural field study.

		F→F			F→T			T→F			T→T			Total		
		D-T	D-D	CI	D-T	D-D	C.I.	D-T	D-D	C.I.	D-T	D-D	C.I.	D-T	D-D	C.I.
March	Aver. (m)	66	26	2.27	179	143	1.30	67	55	1.17	157	109	1.37	148	104	1.36
	SD (±)						0.10	31	21	0.11	110	65	0.38	110	65	0.37
	Min (m)						1.18	34	28	1.04	46	43	1.01	34	28	1.01
	Max (m)						1.42	115	85	1.33	537	296	2.81	537	296	2.81
	n			1			3			5		46				55
April	Aver. (m)				120	101	1.22	90	49	1.69	167	113	1.48	159	108	1.47
	SD (±)						0.13	54	23	0.26	136	87	0.45	136	87	0.43
	Min (m)				45	36	1.11	51	32	1.51	37	30	1.01	37	30	1.01
	Max (m)				222	200	1.37	127	65	1.88	600	366	3.11	600	366	3.11
	n						3			2		29				34
May	Aver. (m)									1.60	172	119	1.45	176	121	1.46
	SD (±)									0.61	147	90	0.44	147	90	0.44
	Min (m)									1.12	39	25	1.02	39	25	1.02
	Max (m)									2.29	977	492	3.31	977	492	3.31
	n									3		70				73
June	Aver. (m)				177	108	1.74	191	137	1.41	185	133	1.40	184	131	1.44
	SD (±)				88	61	0.94	66	49	0.36	132	90	0.53	124	85	0.57
	Min (m)				37	33	1.01	141	83	1.09	30	27	1.00	30	27	1.01
	Max (m)				278	193	3.36	269	199	1.79	637	464	3.34	637	464	3.36
	n						6			4		49				59
July	Aver. (m)	89	78	1.13	130	81	1.63	187	132	1.64	118	88	1.30	128	93	1.38
	SD (±)	21	8	0.14	61	40	0.38	105	107	0.59	88	62	0.37	87	66	0.42
	Min (m)	74	71	1.03	48	25	1.10	54	31	1.13	30	28	1.00	30	25	1.00
	Max (m)	104	83	1.23	190	109	1.98	347	301	2.58	379	281	2.65	379	281	2.65
	n			2			4			5		22				33
August	Aver. (m)				65	48	1.31				149	108	1.40	144	104	1.39
	SD (±)				23	10	0.18				118	89	0.512	116	88	0.50
	Min (m)				49	41	1.19				40	29	1.01	40	29	1.01
	Max (m)				81	55	1.44				445	371	3.14	445	371	3.14
	n						2					29				31
October	Aver. (m)	160	76	2.28	74	49	1.61	91	80	1.10	160	104	1.39	132	87	1.46
	SD (±)	4	40	1.07	29	23	0.90	48	37	0.08	137	79	0.32	114	67	0.57
	Min (m)	157	48	1.53	31	26	1.02	31	31	1.00	35	29	1.00	31	26	1.00
	Max (m)	163	105	3.04	114	90	3.76	148	121	1.21	517	314	1.85	517	314	1.85
	n			2			8			4		21				35
December	Aver. (m)	133	101	1.17	141	136	1.04	116	102	1.12	214	156	1.33	168	129	1.23
	SD (±)	137	82	0.23				50	42	0.09	95	65	0.17	101	65	0.19
	Min (m)	43	43	1.00				59	55	1.06	83	70	1.13	43	43	1.00
	Max (m)	334	220	1.51				147	136	1.22	325	239	1.60	334	239	1.60
	n			4			1			3		7				15
Total	Aver. (m)	122	82	1.53	128	89	1.54	136	102	1.37	165	115	1.41	158	111	1.42
	SD (±)	90	58	0.692	79	63	0.693	94	78	0.391	128	83	0.438	122	81	0.47
	Min (m)	44	26	1.00	31	25	1.01	31	28	1.00	30	24	1.00	30	24	1.00
	Max (m)	334	220	3.04	322	272	3.76	403	359	2.58	977	492	3.34	977	492	3.76
	n			9			25			28		273				335

Abbreviations: F: Foraging patch; T: Feeding/resting tree.
Note: A insect foraging patch is defined as an area where the capuchins spent at least seven observation foraging bouts within an area of 11.7 m. Seven observation bouts is equivalent to the average number of bouts which a capuchin spent on a feeding tree, and 11.7 m is the average diameter of feeding/resting trees in the "small forest."
The distances have substractions of 5.9 m, which is the mean radius of the feeding/resting trees in the "small forest."

Use of nodes and route segments

To examine more specifically how white-faced capuchins represent spatial information, I present data on their use of nodes and route segments. During this study, I identified the nodes used by the capuchins to reach feeding/resting sites. A node is defined as a circular area of 30 m in diameter (706.5 m^2) that was used as a crossing point to intersect re-used routes. A route segment is defined as a path re-used by the capuchins that were bounded by two nodes. It had been argued that nodes contain landmark information used for navigation (Garber 2000, Suárez 2003). A primate forager using nodes for traveling on a large spatial scale is expected to encode spatial information in a route-based spatial representation.

At the EBLS, capuchins used a total of 53 nodes during the 8-month study period (Fig. 5.3). On average the capuchins used a total of 12.8±7.0 different nodes per month, 6.2±4.1 different nodes per day, and on average nodes were crossed 1.25±0.51 times per day (Table 5.12, 5.13, Urbani 2009: Appendix D). Fifty-three percent of nodes (28/53) were traveled through only during one month. A total of 20 nodes were visited between two to four months during the study period (37.7%). The remaining five nodes (9.4%) were used across five to six months during the course of the study (Table 5.14). On average each node was crossed 6.1±2.4 days during the 94 observation days of the natural field study (Urbani 2009: Appendix D). The five most commonly used nodes on average were crossed on 36.8±8.4 days during this study. These nodes were crossed on average 44.0±15.0 times each.

To examine the pattern by which capuchins reused routes, I identified the route segments bounded between nodes. The results indicate that a total of 79 different route segments were on average 86±41 m in length. The average distance of route segments range from 63±29 m in October to up to 120±72 m in July (Table 5.15; significantly different: χ^2=31.19, d.f.=7, p<0.01). On average, the monkeys used 12.4±.8.5 route segments per month. The re-use of route segments was similar throughout each month of the study period (not significantly different: χ^2=3.38, d.f.=7, p>0.05; Table 5.15; Urbani 2009: Appendix D). On average route segments were re-used 3.9 times per month, ranging from two times in December to 4.7 times in April. Of the 79 different route segments used by the capuchins during the eight months of the natural field study, one route segment was reused during four months, three route segments were reused during three months, 14 route segments were reused during two months, and 61 route segments were used during only one month of the study.

Another finding of this study was that the number of route segments traveled differed each month, but the use of nodes was more consistent from month to month. This is supported by the following data. The number of different route segments used per month ranged between 26 in March and two in December. The number of different nodes used per month ranged from between 24 in May and four in December. However, as the study progressed the re-use of nodes ranged between 16.7% and 33% in April and May and reached 92.3% and 100% in October and December, respectively (Table 5.13). Thus, during the final months of the natural field study, nodes that were used during the early months of this study were re-used.

	Day 1	Day 2	Day 3	Day 4	Day 5	Day 6	Day 7	Day 8	Day 9	Day 10	Day 11	Day 12	Average	S.D.
March	5	8	3	5	10	6	8	7	4	4	6	10	6.3	2.3
April	6	5	6	3	3	5	4	4	3	3	5	3	4.2	1.2
May	15	11	13	12	13	13	13	15	15	4	18	10	12.7	3.4
June	11	12	14	9	9	11	12	13	10	9	10	14	11.2	1.9
July	3	2	7	7	3	4	6	3	6	4	4	3	4.3	1.7
August	1	3	5	5	4	5	3	4	3	2	3	2	3.3	1.3
October	10	7	6	3	5	6	4	8	8	5	-	-	6.2	2.1
December	2	1	2	1	2	0	1	1	3	0	2	3	1.5	1.0
Total													6.2	4.1

Table 5.12. Daily use of nodes within the small forest used by members of the La Yunai group during the natural field study.

	Nr. Visited Nodes	Single-used nodes (1st time, one month only)	Reused nodes (≥2 months)	% Reused nodes
March	15	15	0	0
April	6	5	1	16.7
May	24	16	8	33.3
June	21	10	11	52,4
July	10	1	9	90.0
August	9	5	4	44.4
October	13	1	12	92.3
December	4	0	4	100.0
Average	12,8			
SD	7			
Min.	4			
Max.	24			

Table 5.13. Monthly use of nodes within the small forest used by members of the La Yunai group during the natural field study.

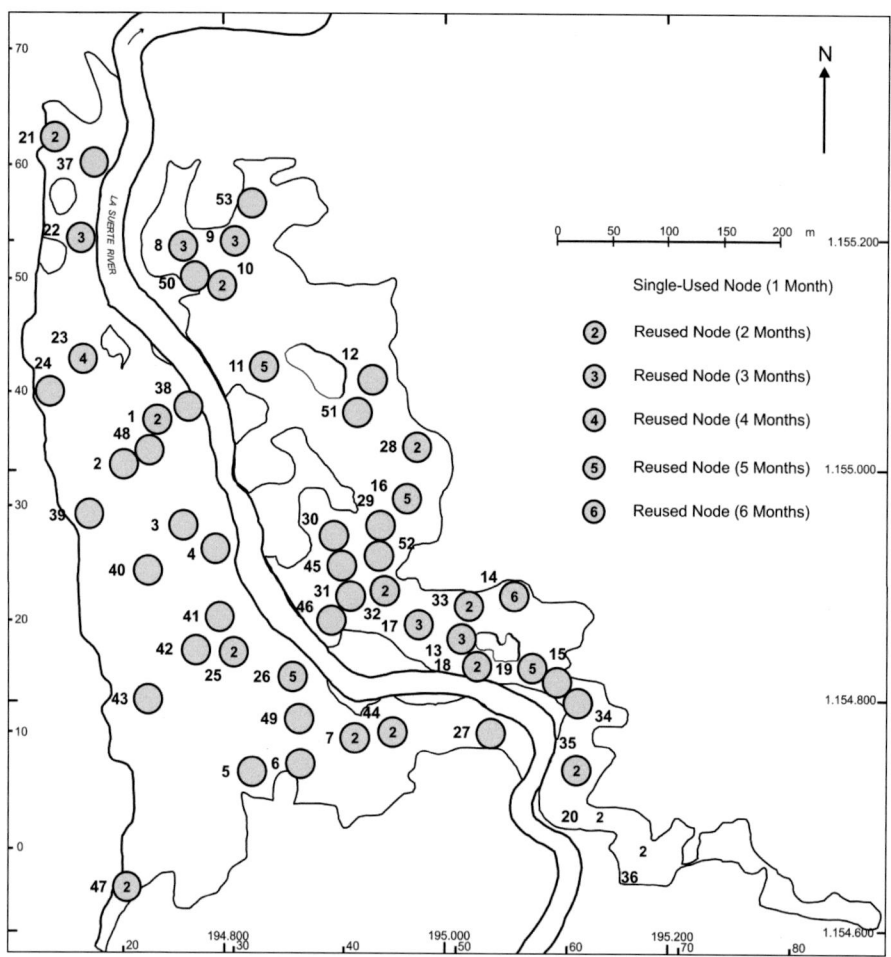

Figure 5.3. Monthly nodes used by the La Yunai group during the natural field study: Annual.

Nr. Months visited	Nr. Nodes	% Nodes Visited
1	28	52.8
2	13	24.5
3	6	11.3
4	1	1.9
5	4	7.5
6	1	1.9

Note: *n*=53 nodes. Each time visited value occurred during one month.

Table 5.14. Pattern of visits to nodes used by members of the La Yunai group during the natural field study.

		March	April	May	June	July	August	October	December	Total
Distance of segments	**Average**	76	102	77	96	120	96	63	105	86
	SD (±)	31	38	38	36	72	29	29	-	41
	Max.	126	153	168	179	247	116	147	116	247
	Min.	42	68	42	37	53	53	37	95	37
	n	14	6	26	23	9	6	13	2	99*
Reuses of segments	**Average**	3.3	4.7	4.3	4.0	4.0	2.0	3.8	2	3.9
	SD (±)	2.1	3.4	2.0	1.6	2.1	0.9	2.2	-	2.0
	Max.	8	11	9	7	7	3	8	2	11
	Min.	1	2	1	1	2	1	1	1	1

*Note: A total of 79 individual route segments were used during the study period. *The total of 99 in the table is due to the fact that some route segments were used in different months of the natural field study.*

Table 5.15. Distances between nodes and re-use of segments used by members of the La Yunai group during the natural field study.

Reaching the major feeding/resting trees from different directions

The ability of an efficient forager to revisit feeding/resting sites from different directions and locations in the forest is consistent with a coordinate-based map. However, if a forager traveled through the same set of nodes and route segments to revisit a feeding site, this is more consistent with a route-based spatial representation. In order to examine the ability of capuchins to reach major feeding/resting trees from different directions or previous feeding/resting sites, I analyzed data on the sequential use of feeding/resting sites and directions of capuchins´ travel. This was accomplished by drawing a straight-line from the previous tree to the target, and measuring the angle of straight-line travel. These angles were represented in azimuth rosettes and grouped in units of 5° to represent arrivals from different cardinal directions.

Overall, each major feeding/resting tree was visited on 18.4±13.4 different occasions during the study period (Table 5.16). The mean number of visits from different directions to major trees was 11.7±7.3 (Directions used one time: 7.7±3.8, Directions used multiple times: 4.0±4.2; Table 5.16). For example, the most frequently visited tree by white-faced capuchin was reached 49 times, and 26 of these were from different directions (feeding Tree #9: *Dipteryx panamensis*). Resting Tree #10 (*Stryphnodendron microstachyum*) was visited 26 times, and 16 of those were from different directions. At the EBLS, all major feeding/resting trees were reached from multiple directions (Table 5.17, Fig. 5.4). These results are consistent with a model of spatial memory in which a forager either possibly associated multiple landmarks with the same target tree or encoded coordinate-based spatial information to locate these feeding/resting trees.

In addition, to distinguish between these alternatives, I examined the frequency with which the capuchins re-visited nodes and route segments to reach major feeding/resting trees. Based on reused of the top five nodes (nodes that were used in five or 6 months), the results indicate that 82.6% of the time the capuchins revisited major feeding/resting trees using the same node. This pattern of use of nodes is consistent with the use of a route-based spatial representation.

	T1	T9	T10	T12	T32	T77	T94	T114	T182	Average	SD (±)
Visits during the study period	27	49	26	11	11	14	8	10	10	18.4	13.4
Visits from different directions	20	26	16	7	6	10	6	6	8	11.7	7.3
a) Direction used one time	15	12	10	5	4	6	5	5	7	7.7	3.8
b) Directions used multiple times	5	14	6	2	2	4	1	1	1	4.0	4.2
Details (Directions used multiple times)											
Directions used 2 times	2	9	5	1	1	4	1	0	1	2.7	2.9
Directions used 3 times	3	3	0	0	0	0	0	0	0	0.7	1.3
Directions used 4 times	0	1	0	1	0	0	0	0	0	0.2	0.4
Directions used 5 times	0	0	0	0	1	0	0	1	0	0.2	0.4
Directions used 6 times	0	1	1	0	0	0	0	0	0	0.2	0.4

Table 5.16. Visits and directions to major feeding/resting trees used by the La Yunai group during the natural field study.

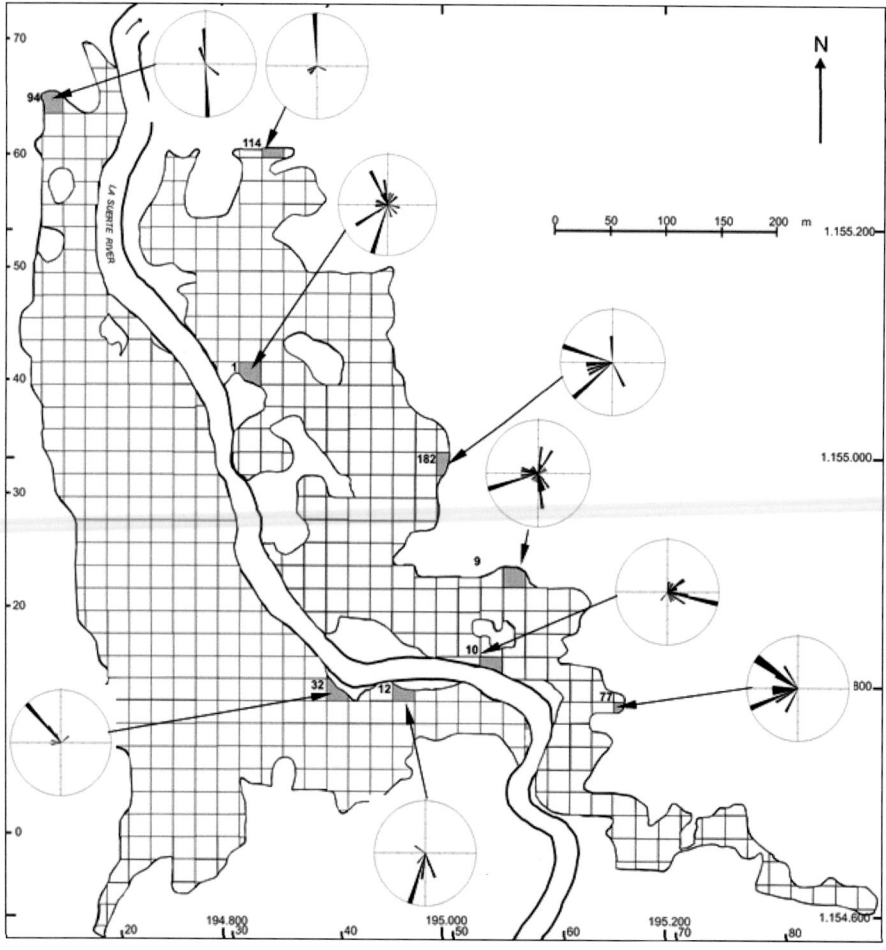

Figure 5.4. Azimuth rosettes around major feeding/resting trees used by the La Yunai group during the natural field study. The angular width of every "ray" is 5° degrees. In terms of length, longer "rays" imply greater arrivals from that given direction. Shorter "rays" indicate a lower number of arrivals from that direction.

Tree #	# of directions	# of 45° portions that surrounded the rosette	d.f.	χ 2	P	Distribution of directions
T1	27	8	7	12.41	>0.01	Homogeneous
T9	27	4	3	3.91	>0.01	Homogeneous
T10	23	5	4	20.70	<0.01	Non-homogeneous
T12	11	5	4	15.82	<0.01	Non-homogeneous
T32	10	5	4	23.00	<0.01	Non-homogeneous
T77	14	5	4	13.14	>0.01	Homogeneous
T94	4	4	3	2.00	>0.01	Homogeneous
T114	5	4	3	3.80	>0.01	Homogeneous
T182	10	5	4	11.00	>0.01	Homogeneous

Note: The maximum number of 45° sections surrounding the rosettes is 8. The cases in which there are fewer than eight sections are due to border constraints (see Fig. 5.2). Thus, the number of 45 ° sections varies.

Table 5.17. Directness of arrivals to major feeding/resting trees used by members of the La Yunai group during the natural field study.

Discussion

A main issue in primate cognitive ecology is how variation in habitat utilization and ranging are influenced by differences in the manner in which primates use spatial, social, and ecological information (Janson and Byrne 2007, Garber *et al.* 2008). Many primate species travel to feeding and resting trees that are far outside field-of-view (Milton 1980, Boesch and Boesch 1984, Garber 1989, 2000, Janson 1996, 1998, Di Bitetti and Janson 2001, Garber and Jelinek. 2005, Di Fiori and Suárez 2007, Noser and Byrne 2007, Normand and Boesch 2009). However, the way in which primates encode, store, and hierarchically integrate information remains unclear (Garber 2000, Janson 2000, Cunningham and Janson 2007, Garber *et al.* 2008).

In this chapter, I examined if capuchin spatial representation and decision-making are more consistent with a route-based map or a coordinate-based map (Byrne 1982, 2000, Poucet 1993, Garber 2000, Milton 200, Suárez 2003, Di Fiori and Suárez 2007). Using a route-based spatial representation a forager is expected to encode route segments and uses landmarks or nodes to orient travel in order to locate targets or goals. In contrast, a forager using a coordinate-based spatial representation stores information as X and Y coordinates, encoding exact angles and distances, and computing novel travel routes to reach goals (Byrne 1979, 2000; Bennet 1996; Cheng and Spetch 1998; Garber 2000; Suárez 2003).

Poucet (1993) has offered an integrative model to hypothesize how a forager might represent information in both a larger spatial scale (e.g. between distant feeding sites) and a smaller spatial scale (e. g. vicinity of a feeding site). He argues that in larger scale space, foragers are expected to use route-based information relying on landmarks as stable features of the environment to guide travel. These landmarks are expected to be located or visible at nodes or cross points and are reached by re-using a set of familiar route segments. In smaller scale space, however, Poucet (1993) argues that a forager may maintain spatial information in a coordinate-based framework. This is based on the fact that an arboreal forager's ability to obtain and integrate multiple views of goals from different points of the home range and convert this into an internal coordinate-based spatial representation is only possible across a small spatial area due to visual barriers. At present, it remains unclear if primate foragers use a coordinate-based spatial representation, a route-based spatial representation, or a combination of both. However, Normand and Boesch (2009) offer insight into this discussion. Based on 217 observation days on adult individuals, Normand and Boesch (2009) report that chimpanzees (*Pan troglodytes*) living in a western African tropical rainforest tend to use direct travel paths between distant feeding trees. The authors found that chimpanzee route segments deviated only 4% from the most direct route between feeding sites. Moreover, Normand and Boesch (2009) found that such a pattern of high linearity (linearity index of 0.96/1. Direct distance/distance traveled: 0 is backtracking, 1 is straight-line progression. A linearity index of 0.96 is equivalent to a circuity index of 1.04) was consistent across the entire home range of the chimpanzees. In addition, the chimpanzees reached the same feeding trees using different previous directions. Given these results, the authors conclude that chimpanzees appear to maintain accurate notions of directions and distances between widely scattered trees across their home range and use a coordinate-based spatial representation for navigating between previous visited feeding sites.

In 1934, R. Carpenter first indicated that wild white-faced capuchins (*Cebus imitator*) in Panama use a goal-directed movement pattern in locating feeding trees. Carpenter defined the goal-directed movement as the use of an interconnected network of "pathways" that are re-used and learned in order to reach "preferred" goals (Carpenter 1934). Kühlorn (1939) reported that in Mato Grosso-Brazil, wild black-horned capuchins (*Sapajus cirrifer*; today *S. nigritus*) reused habitual routes of travel within the forest. The data Kühlorn used to support his contention was collected *ad libitum* after observing the recurrent use of pathways by the capuchins. In his pioneering research on the behavior and ecology of Venezuelan wedged-capped capuchins (*C. olivaceus*), Robinson (1986) indicated that over the course of a year, capuchins revisited a set of feeding trees using the same set of travel pathways.

Robinson (1986) reported that capuchins re-used travel segments and visited feeding trees from multiple directions. He found that capuchins tended to move forward and reduced backtracking to reach "feeding patches." In order to test if wedge-capped capuchins showed evidence of mental mapping of their environment, Robinson (1986) evaluated the manner in which they visited the most commonly used feeding trees. Based on 99 days of observation, he found that the capuchins tracked changes in food availability across their home range and revisited particular fruit-feeding sites as well as water ponds. Robinson (1986) concluded that capuchin travel is positively correlated to resource distribution and availability rather than intergroup contacts. In addition, he pointed out that the capuchins "have stable ranges, well-traveled pathways, and habitual sleeping and drinking sites" (Robinson 1986: 40), and that they appear to have detailed spatial knowledge of the location of feeding sites within their home range. These results appear to suggest that wedge-capped capuchins maintained a route-based spatial representation of their home range.

At the EBLS, white-faced capuchins visited 10 feeding/resting trees each day. Two of these trees on average were revisited the same day. Revisits to the same tree on the same day suggest that capuchins exploit large and highly productive feeding sites. One-fourth of the feeding trees fed in by white-faced capuchins each month was used for the first time. During 62 of the 94 observation days of this study, white-faced capuchins visited at least two trees of the same species. These results suggest that during feeding and traveling white-faced capuchins monitored the availability and phenology of new feeding sites and encoded this information into a mental map of the spatial distribution of feeding sites. It also is possible that some feeding sites are encountered by chance as individuals travel between previously visited feeding sites. I describe one anecdotal example of possible resource monitoring. On March 6th at 10:12 am, the focal animal was observed to manually and visually inspect the fruits of an *Inga spectabilis* tree, while two other capuchins were similarly inspecting and biting (but not eating) the fruits. At that time, the *I. spectabilis* fruits were unripe. One month later (April 6th, 7:10 am), the fruits had ripened and the tree was re-visited by the capuchins and the fruit consumed. A total of four *I. spectabilis* trees were visited by the capuchins during that same month. In sum, given that visibility in the canopy was generally <25 m, it is likely that the capuchins integrated phenological information on fruiting synchrony with spatial information on tree species distribution to select feeding sites.

In the Peruvian Amazon, Garber (1988) indicated that when exploiting the nectar of *Symphonia globulifera*, saddle-back and moustached tamarins (*Leontocebus fuscicollis* and *Saguinus mystax*) selected the nearest-neighbor trees of this species. A total of 10 *S. globulifera* trees were located in the troop´s home range and the mean nearest-neighbor distance between these trees was 70.4±68m. Tamarins visited a total of 5.1 individual trees of this species per day (range: five to eight trees) and spent a mean time of 43.9 minutes per day consuming nectar and handling flowers of *Symphonia* (range: 14 to 74 min). Garber (1988) reported that tamarins selected the nearest flowering tree as the next feeding site in 86% of feeding bouts (n=533 for *L. fuscicollis*, n=467 for *S. mystax*). He concluded that the tamarins remembered the location of these trees within their home range, and selected *Symphonia* trees based principally on minimizing travel distances between feeding sites.

Based on the second dataset of three months, Garber (1989) expanded this study to examine tamarin foraging decisions when exploiting 150 individual trees of 20 tree species. These trees accounted for ~75% of tamarins´ feeding time. Trees of individual species were highly synchronized in their fruiting phenology. Tamarins used on average 5.8±2.2 of those trees per day (range: 4–11 individual trees). When visiting these trees, the tamarins used relatively straight-line travel itineraries (~150 m in length) and reached these major feeding trees from several different directions. Garber (1989) suggested that tamarins remembered the location of a large number of trees in their home range and select feeding sites using a distance-minimizing travel itinerary. White-faced capuchins at the EBLS also were found to select the nearest feeding tree when exploiting two tree species *Dipteryx panamensis* and *Sterculia recordiana*. These data are consistent with the hypothesis that the capuchins maintain detailed knowledge of the location of different feeding/resting sites within the forest.

In a study in the Ecuadorian Amazon, Di Fiori and Suárez (2007) report that woolly and spider monkeys (*Lagothrix lagotricha* and *Ateles belzebuth*) repeatedly reused the same routes of travel across their 400 ha home range. Suárez (2003) found that spider monkeys relied on intersection points or nodes as "places of choice" where routes interconnect when navigating between feeding sites. Di Fiori and Suárez (2007) indicated that the movement patterns of woolly monkeys and spider monkeys were most consistent with a route-based map.

Capuchins at the EBLS were found to re-use route segments and nodes within their home range. In addition, major feeding/resting trees were visited from multiple directions. Overall, the use of particular route segments and nodes varied from month to month; however, as the study progressed the re-use of nodes was more consistent. I found that 67.2% of the time when they revisited the major feeding/resting trees from different directions, the capuchins traveled through and oriented to the same nodes. There were no physical barriers between nodes and the major feeding/resting trees. The reuse of travel segments and nodes appeared to be consistent with a route-based spatial representation.

Although capuchins often selected relatively direct routes between feeding/resting sites, there was little evidence of the use of novel shortcuts which is expected using a coordinate-based map. Overall the mean distance to targets was 111±81 m, and the capuchins' travel deviated 42% more than the straight-line distance. However, at approximately half the distance (55±5 m), the capuchins traveled only 14% more than the straight-line itinerary, and at a distance of 30 m around the perimeter of the feeding/resting trees, the capuchins' travel deviation was 7%. A similar pattern of more direct travel was reported by Garber (2000) in mustached and saddleback tamarins when traveling in the vicinity (50 m) from major feeding sites. In a western African rainforest, Normand and Boesch (2009) argued that the ability of chimpanzees to reach feeding trees from a variety of different straight-line directions (CI=1.04) is consistent with a coordinate-based spatial representation. Thus, it appears that in large-scale space chimpanzees may be relying on a different type of spatial representation than capuchins and tamarins. In smaller scale space, however, these neotropical primates appear to form coordinate-based spatial representations.

In conclusion, capuchins use a route-based spatial representation while traveling within their home range (large-scale space). Data such as the of use a relatively straight-line progression once they reach the vicinity of feeding/resting trees (30 m) and reach these trees from multiple directions, suggest that in small-scale space white-faced capuchins navigate using a coordinate-based spatial representation. This pattern is consistent with the model of spatial representation formulated by Poucet (1993).

However, the capuchins also exhibit ranging patterns that are not consistent with the expectations of coordinated-based spatial navigation. This includes, for example, high deviation in travel itineraries and the use of nodes for navigation. Although the use of novel shortcuts to reach nearby feeding/resting sites serves to reduce travel time and travel distance, by taking relatively direct but alternative routes, a forager may benefit by sampling information on food availability, fruit ripeness, and insect distribution. If this is the case, then we need to reconsider Poucet's model and include the requirements of resource sampling in primate decision-making. In this regard, ranging patterns and travel routes reflect both the sensory/cognitive ability of the forager as well as the set of ecological problems foragers encounter, such as locating resources, and the most efficient ways of solving these problems.

6

Spatial Mapping in Wild White-Faced Capuchin Monkeys (*Cebus imitator*): An Experimental Field Study

In this chapter, I present data based on a set of controlled experimental field studies designed to examine spatial representations in white-faced capuchin monkeys (*Cebus imitator*). In particular, I test whether capuchins have route-based or coordinate-based spatial memory when visiting feeding platforms. In these experiments, capuchins using a route-based representation are expected to have higher circuity indices and to re-use traditional travel routes when moving between experimental platforms. In contrast, capuchins using a coordinate-based spatial representation are expected to have lower circuity indices and use novel shortcuts and direct routes of travel when moving between experimental platforms.

The specific questions I examine are: (a) do white-faced capuchins visit experimental feeding platforms in the same order as they were first encountered?, (b) do white-faced capuchins travel between sequential feeding platforms using a distance minimizing principle, (c) over the course of the field experiments do the capuchins reduce their circuity index in traveling between experimental platforms?, (d) do white-faced capuchins use traditional routes of travel to reach experimental feeding platforms or do they select novel travel routes?, (e) do white-faced capuchins select nearer platforms over more distant platforms?, (f) do white-faced capuchins preferentially travel to more distant feeding platforms that contain higher food rewards over nearer feeding platforms that contain lower food rewards?, and (g) do white-faced capuchins exhibit a win-shift foraging rule when selecting experimental feeding platforms?.

Material and Methods

During a 3-month period, two field experiments were conducted (September to November). Experiment 1 (Equal distance/Equal amount of food reward) lasted 24 consecutive days, including five days of pre-baiting. The objective of this experimental design was to examine the routes taken by white-faced capuchins when traveling to equally distant feeding sites of equal food quantity (eight bananas each). The platforms were located at

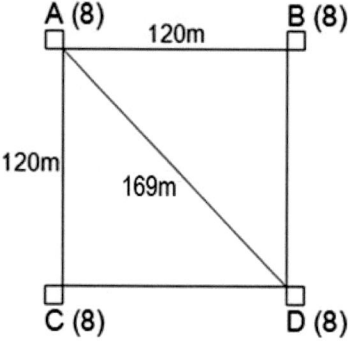

Figure 6.1. Experiment 1: Equal distance/ Place constant/Equal amount of food reward (Vertices and letters: Platforms and their codes; in parentheses: Amount of food reward).

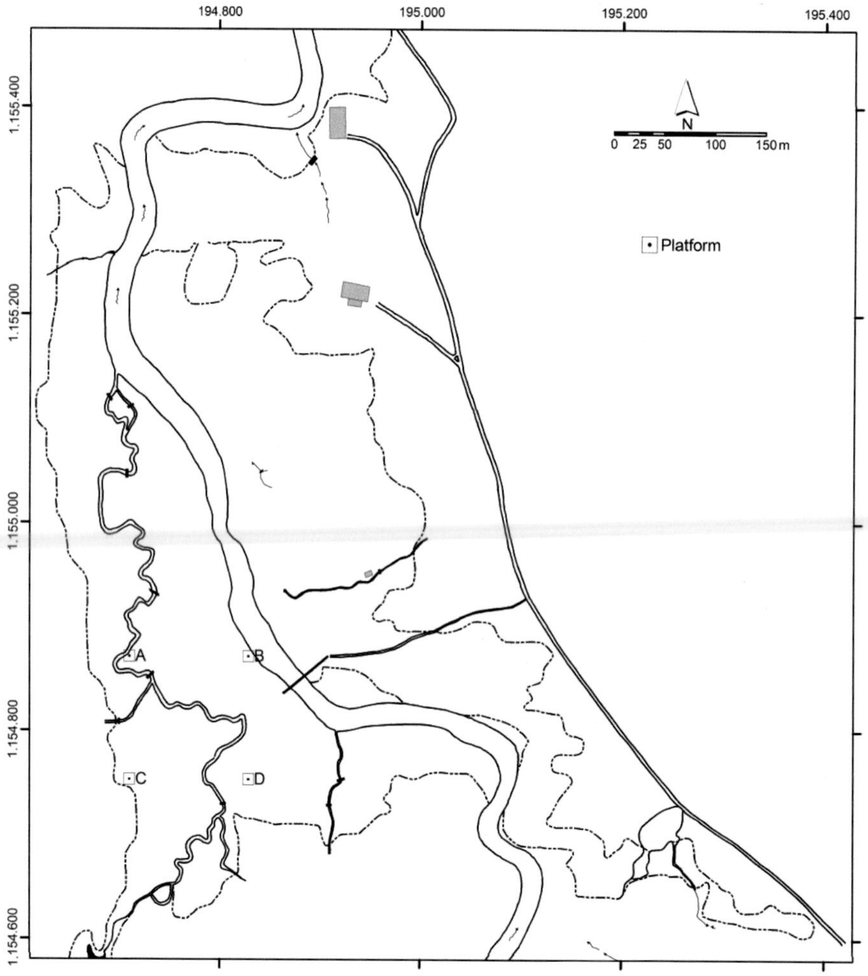

Figure 6.2. Location of platforms of Experiment 1 within the EBLS.

NAVIGATING WITH WHITE-FACED CAPUCHIN MONKEYS

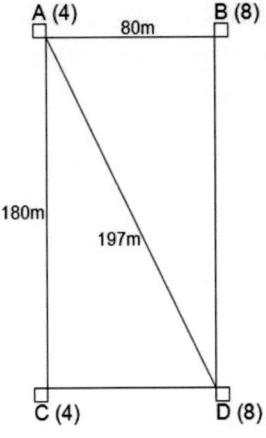

Figure 6.3. Experiment 2: Different distances/Place constant/Higher *vs.* lower food reward (Vertices and letters: Platforms and their codes; in parentheses: Amount of food reward).

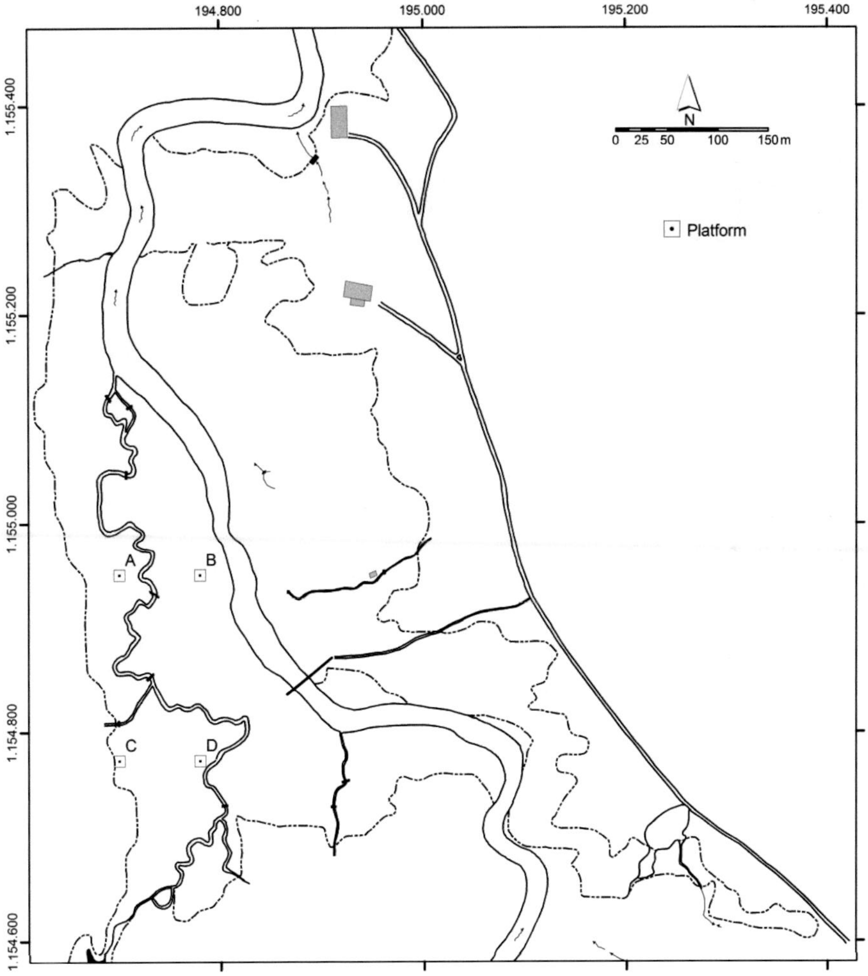

Figure 6.4. Location of platforms of Experiment 2 within the EBLS.

a distance of 120 m from each other, and at a distance of 169 m from opposite vertices (Fig. 6.1, Fig. 6.2). The distance of 120 m was selected because it is similar to the mean distance between naturally occurring capuchin feeding/resting trees recorded from March to August (111.7±14.8 m).

Experiment 2 started 12 days after the completion of Experiment 1. Experiment 2 (Different distances/Higher *vs.* lower food reward) was conducted during 43 consecutive days, including a 12 pre-baiting day period. The objective of this experimental design was to examine the ability of white-faced capuchins to make foraging decisions that require tradeoffs between quantity and distance information. That is; did white-faced capuchins preferentially select nearer and less productive feeding sites or did they preferentially select more distant but more productive feeding sites? In this experiment, two platforms each contained four bananas and two platforms each contained eight bananas. The platforms were located in a rectangular configuration with two sides of 80 m and two sides of 180 m. The distance between opposite vertices was 197 m (Fig. 6.3, Fig. 6.4). In both experiments platforms were baited once per day.

Results

Experiment 1
Do white-faced capuchins visit experimental feeding platforms in the same order as they were first encountered?
In order to examine whether white-faced capuchins visit feeding platforms in the same order they were first encountered, I compared sequences of platform visits across days. On Day 1 capuchins first discovered Platform 1B. On Day 2, Platforms 1D and 1C were discovered in this order. Platform 1A was first visited on Day 5. During the 24-day experiment, the capuchins visited an average of 3.5±1.5 platforms per day. Overall, Platform 1A was visited 12 times (20.3%). Platform 1B was visited 20 times (33.9%). Platform 1C was visited 14 times (23.7%). Platform 1D was visited 13 times (22%) (Table 6.1). So, 25% was expected for each platform.

Platform 1B was selected first by the capuchins on 87.5% (14/16) of the sample days when at least two platforms were visited in sequence (Table 6.1). On those days in which Platform 1B was visited first, Platform 1D was selected second 56.3% (9/16) of the time. Platform 1C was selected second 31.3% of the time (5/16) and Platform 1A was selected second 12.5% of the time (2/16). These results suggest that the initial order of platform discovery played an important role in the capuchins platform or patch choice in this experiment.

The most common pattern of platform visit used by the capuchins was B→D→C. The white-faced capuchins visited the platforms in this way on seven of 17 occasions. On 14 days, Platform 1B was the first platform visited 18 times, and on three days, Platform 1A was visited the first three times. Platform 1D was visited first one time, and Platform 1C was never the first platform visited by the capuchins. In four days the capuchins visited all four experimental platforms.

In Experiment 1, I also examined whether or not capuchins were more likely to find a feeding platform if it was located nearby to a quadrat commonly traveled through during the natural field study (March to August). This was accomplished by drawing a radius

Experimental days	Time of visit	1st visit	2nd visit	3rt visit	4th visit (if any)	5th visit (if any)	6th visit (if any)	Step-by-step movement	Back-tracking movement	Comments
Day 1										B was found (AM: B→B)
Day 2										D, C were found. B was also visited. (AM: B→ D →B→C)
Day 3										B, D were visited. (AM: B→D→B→D→D)
Day 4										B, C, D were visited. (AM: B→D→D→C; PM: D)
Day 5										A was found. B, C, D were also visited. (AM: B→B→D; PM: D→C→A)
Day 6	AM	B→	D→	C→	A	-	-	X	-	Visit sequence counting
	PM	A	-	-	-	-	-	-	-	begins on Day 6, once
Day 7	AM	B→	D→	C→	A	-	-	X	-	all four platforms were
	PM	-	-	-	-	-	-	-	-	visited and known.
Day 8	AM	B→	A	-	-	-	-	-	-	
	PM	B→	D	-	-	-	-	-	-	
Day 9	AM	B→	D→	C→	A	-	-	X	-	
	PM	-	-	-	-	-	-	-	-	
Day 10	AM	B→	D→	C	-	-	-	X	-	
	PM	B→	C→	A	-	-	-	X	-	
Day 11	AM	B→	D→	C	-	-	-	X	-	
	PM	-	-	-	-	-	-	-	-	
Day 12	AM	-	-	-	-	-	-	-	-	
	PM	-	-	-	-	-	-	-	-	
Day 13	AM	B→	C→	A	-	-	-	X	-	
	PM	-	-	-	-	-	-	-	-	
Day 14	AM	B→	C→	D	-	-	-	X	-	
	PM	B	-	-	-	-	-	-	-	
Day 15	AM	B→*	D→*	A→*	C*	D→**	B**	X*	-	*=1st focal monkey
	PM	-	-	-	-	-	-	-	-	**=2nd focal monkey
Day 16	AM	D→	C→	B	-	-	-	X	-	
	PM	-	-	-	-	-	-	-	-	
Day 17	AM	B	-	-	-	-	-	-	-	
	PM	-	-	-	-	-	-	-	-	
Day 18	AM	B→	D→	C	-	-	-	X	-	
	PM	-	-	-	-	-	-	-	-	
Day 19	AM	A→	C→	A→	C→	D		-	X	
	PM	-	-	-	-	-	-	-	-	

Table 6.1. Foraging pattern in Experiment 1: Equal distance constant/Place constant/Equal amount of food reward.

Experimental days	Time of visit	1st visit	2nd visit	3rt visit	4th visit (if any)	5th visit (if any)	6th visit (if any)	Step-by-step movement	Back-tracking movement	Comments
Day 20	AM	-	-	-	-	-	-	-	-	
	PM	-	-	-	-	-	-	-	-	
Day 21	AM	A	-	-	-	-	-	-	-	
	PM	-	-	-	-	-	-	-	-	
Day 22	AM	B	-	-	-	-	-	-	-	
	PM	B→	A	-	-	-	-	-	-	
Day 23	AM	B→	D→	C	-	-	-	X	-	
	PM	-	-	-	-	-	-	-	-	
Day 24	AM	B	-	-	-	-	-	-	-	
	PM	-	-	-	-	-	-	-	-	
Total								12 (92.3%)	1 (7.7%)	= number of platform sequences (percentage).

Abbreviations: Each platform was designated with a capital letter from A to D; →= from platform "X" to platform "Y;" AM= Ante meridian (from time of departure from sleeping site to 11:59); PM= Post meridian (from 12:00 to sleeping site arrival time). *Note*: All platforms had eight bananas.

Table 6.1. continued.

Platform ID	Total grid points inside the forest	Unused grid points inside the forest	Used grid points inside the forest	% Used grid points inside the forest	Order of discovery of platforms
A	87	27	60	69.0 (60 out of 87)	4th discovered
B	68	11	57	83.8 (57 out of 68)	1st discovered
C	64	12	52	81.3 (52 out of 64)	2nd discovered
D	94	27	67	71.1 (67 out of 94)	3rd discovered

Note: n=109 (maximum number of grid points in each 60 m radius).

Table 6.2. Comparison of the grid points used during the natural field study within a radius of 60 m in relation to the discovery of the feeding platforms used by the La Yunai group during Experiment 1.

of 60 m around each platform (half the distance of the 120 m x 120 m square configuration of this experiment) and calculating the number of times capuchins had previously visited that area (grid points). I found that the order of platform discoveries was positively related to the platform's proximity to areas of the forest commonly visited (Figure 6.5, Table 6.2). For example, 83.8% (57/68) of the 10 × 10 m grid points located within a 60 m radius of Platform 1B had previously been visited by the capuchins. In the case of Platform C, 81.3% of the surrounding grid points had previously been visited. The data for Platforms 1A and 1D indicate that capuchins traveled through 69% (Platform 1A) and 71.1% (Platform 1D) of these grid points.

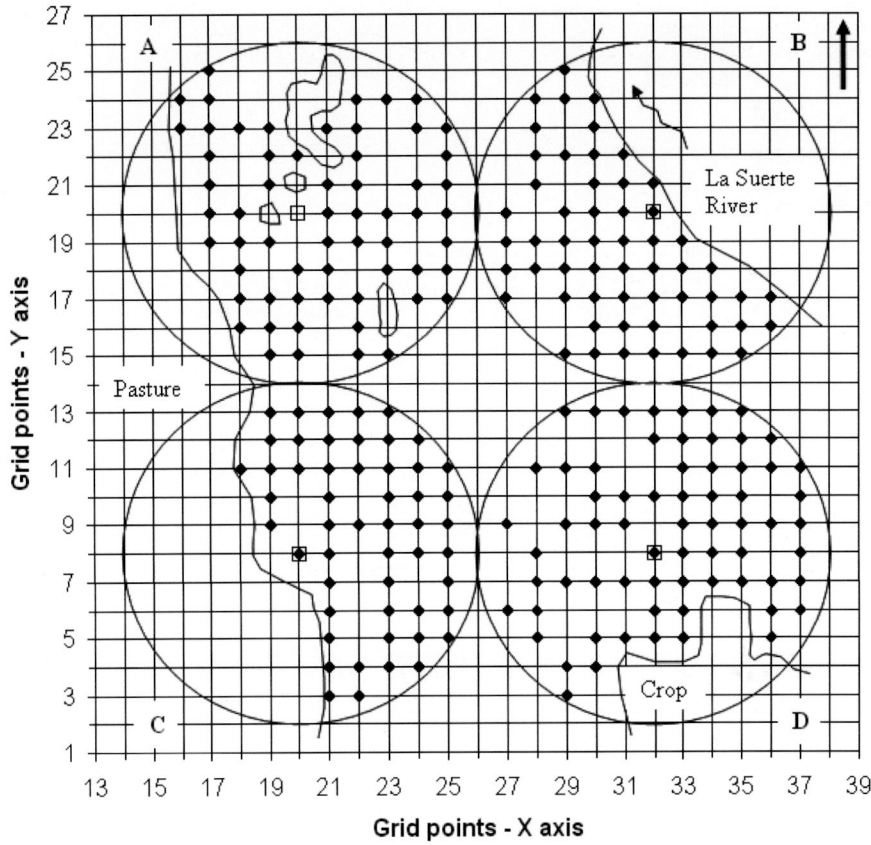

Figure 6.5. Distribution of grid points used during the natural field study within a radius of 60 m in relation to the location of the feeding platforms used by the La Yunai group during Experiment 1 *Notes*: 1) Black dots are used grid points during the natural field study; 2) Empty squares at the middle of the circles are feeding platforms; 3) The scale (or space) between the black dots is 10 m.

Do white-faced capuchins travel between sequential feeding platforms using a distance-minimizing principle, and over the course of the field experiment do they reduce their circuity index in traveling between experimental platforms?

To test if capuchins traveled directly between experimental feeding platforms, I calculated the index of circuity between feeding platforms that were visited in succession. The capuchins visited two platforms in succession four times, three platforms in succession eight times, and four platforms in succession five times. In Experiment 1 (square configuration) with platforms located a distance of 120/169 m apart, capuchins traveled 24% more than the most direct distance to reach the platforms (CI=1.24). In 22.1% of cases CI was <1.10 (5/22), and in all instances, this involved traveling from Platform 1B to Platform 1D (distance: 120 m). In 40.9% of cases, CI ranged between 1.10 and ≤1.25 (9/22) and in 36.5% of the cases CI was >1.25 (8/22). I also calculate the CI when the capuchins traveled the final 60/85 m to reach a feeding platform (half the distance between platforms). At a distance of 60/85 m mean CI was 1.09, with 77.3% of the cases having a CI value of <1.10 (17/22). At a distance of 30 m, the capuchins traveled directly to the feeding platforms exhibiting a CI of 1.03.

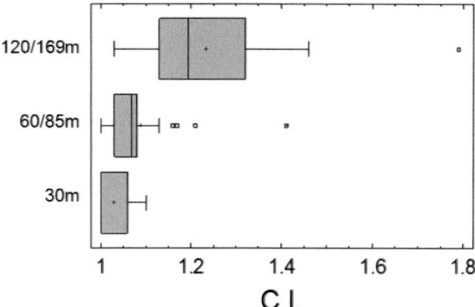

Figure 6.6. Comparison of the circuity indices of Experiment 1.

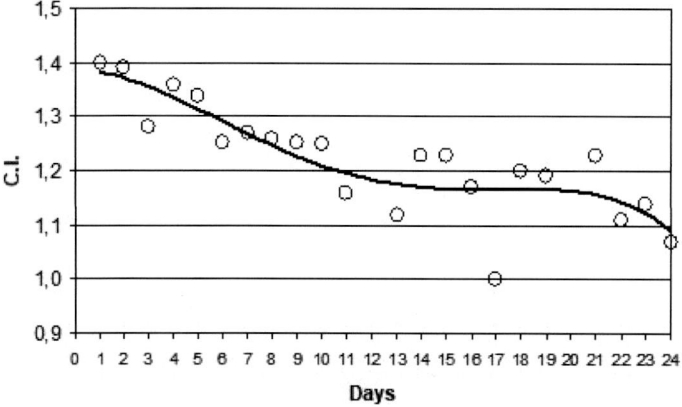

Figure 6.7. Learning curve with the circuity indices of Experiment 1 (*Note*: Fourth-degree polynomial line, $R^2=0.7168$).

To determine if the capuchins used more direct routes of travel as they approached their goal, I compared the circuity indices at 120/169 m, 60/85 m, and 30 m. In all cases, there were significant differences between the mean CI values (120/169 m *vs.* 60/85 m: $Z=3.53$, d.f.=42, $p<0.01$; 120/169 m *vs.* 30 m: $Z=5.56$, d.f.=42, $p<0.01$; 60/85 m *vs.* 30 m: $Z=2.94$, d.f.=42, $p<0.01$). Fig. 6.6 summarizes this pattern. Overall, the capuchins appeared to adjust their progression and use a more direct travel route as they approached their goal even at distances in which they could not sight directly to the target (see below).

To identify if capuchins encoded and incorporated spatial information over time to improve their travel efficiency, I present a learning curve with the CI values calculated in Experiment 1 (Fig. 6.7). For example, at the beginning of the experiment (Day 1 to 5), the capuchins exhibited a CI that ranged between ~1.3 and ~1.4. These CI values decreased over time and by Day 12 approach 1.10 and remained at that level for the remaining of the experiment. To examine whether CI values decreased during the course of the experiment, I compare CI values on Days 1–5 with Days 20–24. The results were significantly different (CI Days 1–5=~1.4-~1.3; CI Days 20–24=~1.17-~1.1; $t=2.61$, d.f.=21, $p<0.01$), and suggest that the capuchins decreased travel distance over time by learning to use more efficient routes between experimental platforms.

Do white-faced capuchins use traditional routes of travel to reach experimental feeding platforms or do they select novel travel routes?

To examine if capuchins formed novel travel routes after initially finding new feeding sites (Urbani 2009: Appendix E), or whether they continued to use traditional routes of travel, I calculated the degree to which routes used by the monkeys during the natural field study also were used during Experiment 1 (March to August). The area of the square configuration of platforms (120 m x 120 m) was 1.4 ha. Within this area, the overlap between routes used during the natural field study and the routes used during this experiment was only 4.3%. Travel routes were contained in 65.8% of the area inside the perimeter formed by the experimental platforms during the course of this experiment. I never observed the capuchins using these routes during the natural field observation (March-August). The remaining 30% of the area was not used during the natural field study or during the field experiment.

A total of 36 quadrats (20 m x 20 m) were represented within the configuration of Experiment 1. Twenty-three quadrats (63.9%) were crossed by new routes traveled during this experiment. Five quadrats (13.9%) contained routes that were used by capuchins during both the natural field study and the experimental field study. The remaining eight quadrats (22.2%) were not used. The results indicate that capuchins formed and used new routes when traveling between experimental feeding platforms; however, these routes did not represent novel shortcuts or straight-line travel between feeding platforms. This is consistent with spatial information in the form of a route-based map.

Four areas used as nodes by the capuchins during the natural field study (March-August) were located within the 1.4 ha area bounded by the experimental platforms. These are Node #25 used in May, and nodes #41, #42, and #43 used in June (see Urbani 2009: Appendix D). The four nodes have visited an average of 5.5±1.7 days during the 24-day experiment period, with capuchins visiting 1.0±1.1 nodes per day. These four nodes were crossed on 22 occasions. (Node #25=7, Node #41=6, Node #42=6, and Node #43=3; Table 6.3). These results strongly support the contention that capuchins use a route-based spatial representation of large-scale space.

To examine the patterns by which capuchins used routes, I identified five route segments bounded by these nodes in this experiment. The average length of the route segments was 53.2±18.3 m. The average distance of route segments ranged from 84.9 on Day 5 to as low as 37.7 m on Day 15 (Table 6.4). I found no cases of capuchins re-using route segments between these nodes during this experiment. However, the route segments formed during this experiment bounded between nodes #41 and #42, and nodes #42 and #43 were also used during the natural field study. In June, the route segment bounded between nodes #41-#42 was used on three days, and the route segment bounded between nodes #42-#43 on two days. The use of habitual route segments is found in route-based spatial representations. The results also suggest that capuchins formed new routes when traveling between experimental feeding platforms; however, the novel route segments did not represent shortcuts between feeding sites. In sum, this is consistent with spatial information encoded as a route-based system.

Days:	1	2	3	4	5	6	7	8	9	10	11	13	14	15	16	17	18	19	21	22	23	24	Total	%
Node 25	1		1	1						1			1	1						1			7	31.8
Node 41							1		1		1		1	1					1				6	27.3
Node 42			1										1	1			1	1	1				6	27.3
Node 43				1										1			1						3	13.6
	1	0	2	2	0	0	1	0	1	1	1	0	3	4	0	0	2	1	2	1	0	0	22	

Table 6.3. Nodes recorded during the natural field study used in Experiment 1.

	Node 25	Node 41	Node 42	Node 43
Node		Day 5	Day 3	Day 5
25		(47.2 m)	(49.1 m)	(84.9 m)
Node			Day 15	
41			(37.7 m)	
Node				Day 15
42				(47.2 m)
Segment length average (m)	53.2			
SD		18.3		

Table 6.4. Route segments recorded during the natural field study used in Experiment 1.

Do white-faced capuchins select nearer platforms over more distant platforms?

In Experiment 1, the distance between all platforms located at the vertices of the square configuration was 120 m, and the distance between diagonally located platforms was 169 m. In order to examine how frequently white-faced capuchins traveled to the more distant platforms, despite the fact that nearer platforms contained an equal food reward, I calculated the number of times after visiting a first platform the capuchins traveled to a nearer or more distant platform. Overall, the probability of selecting a more distant feeding platform randomly after the first platform was visited is one of three or 33%. In this situation, the closer platforms were visited on 88.2% (15/17) of the occasions, while the more distant platform was visited on 11.8% (2/17) of the occasions. The pattern of visit to the closest platforms was significantly greater than chance (χ^2=9.94, d.f.=1, $p<0.01$). Once two platforms were visited the chance of randomly visiting each of the remaining two platforms (unless the first two platforms visited were A→D or B→C, and this never occurred) was 50% each. In this case, the closer platform was selected 83.3% of the time (10/12), while the more distant platform was visited on 16.7% [2/12] of the occasions. Again, the pattern of visit to the closest platform was significantly greater than expected by chance (χ^2=5.33, d.f.=1, $p<0.05$). These results indicate that when the monkeys visited two or more platforms, in general capuchins preferentially selected nearer feeding sites over more distant platforms of equal food reward. This is consonant with the contention that capuchins encode information concerning the distances and spatial relationships between several feeding sites, and use this information in foraging decisions (see similar results in Chapter 5).

Do white-faced capuchins exhibit a win-shift foraging rule when selecting experimental feeding platforms?

I examined if capuchins formed a foraging pattern by visiting a set of platforms in sequence. A sequence was defined as visitation pattern in which the focal animal inspected three or four platforms in succession (see Chapter 3). In this experiment, the capuchins were observed to backtrack, defined as returning to previously visited platform, on only one occasion (1/13) (Table 6.1). These results indicate that capuchins followed a searching strategy based on a win-shift rule in which once a feeding site was exploited, it was no longer part of the set of feeding choices for that day (AM, PM). This type of foraging pattern is associated with traplining. During traplining, a forager visited several feeding sites of the same plant species over the course of a day, and does so in a manner that minimizes backtracking (Janson *et al.* 1981, Garber 2000, Milton 2000). Navigation using traplining assumes that (a) feeding trees of a given species are fruiting at the same time, and (b) movement of the forager is goal-directed between feeding sites.

Experiment 2
Do white-faced capuchins visit experimental feeding platforms in the same order as they were first encountered?

To examine if white-faced capuchins visited feeding platforms in the same order as they were first encountered in Experiment 2, I compared sequences of platform visits during this experiment. As in Experiment 1, the capuchins visited the platforms in the same order as they were first encountered. In Experiment 2, on Day 1 capuchins found Platform 2B and then Platform 2D. Platform 2C was discovered on Day 5. On Day 10, Platforms 2A was first visited. The most common pattern of platform visit used by the capuchins was B→A (Table 6.5). The monkeys visited these platforms in this way on 5 of 19 occasions. During Experiment 2, the capuchins visited an average of 2.6±1.2 platforms per day, less than during Experiment 1. Across the 43 days of this experiment, the capuchins visited Platform 2A 12 times, Platform 2B 20 times, Platform 2C 14 times, and Platform 2D on 13 occasions (Table 6.5).

In Experiment 2, Platform 2B was selected as the first platform visited in 11 of 19 (57.9%) cases in which two or more platforms were visited in succession (Table 6.5). During the days in which Platform 2B was visited first, Platform 2A was selected as the second platform six of 11 times (54.5%). Platform 2D was chosen second 27.3% of the time (3/11), and Platform 2C was chosen second 9.9% of the time (1/11). On one occasion Platform 2B was revisited (9.1%; 1/11). In 15 days, Platform 2B was the first platform visited 19 times (four times when the platform was no longer baited because the capuchins had fed there earlier). In nine days Platform 2D was first. Platform 2C was never the first platform visited by the capuchins. As in Experiment 1, these findings suggest that the monkeys visited the experimental feeding sites in part, based on the order in which these sites were initially discovered.

Experimental days	Time of visit	1st visit	2nd visit	3rt visit (if any)	4th visit (if any)	5th visit (if any)	Short distance/ Less reward	Long distance/ More reward	Step-by-step movement	Back-tracking movement	Comments
Day 1											B, D were found. (AM: B→D)
Day 2											No visit.
Day 3											(AM: B)
Day 4											(AM: D)
Day 5											C was found. (AM: B→D→D→C)
Day 6											(AM: B)
Day 7											(AM: B→C→D)
Day 8											(AM: D)
Day 9											(AM: D→C→B)
Day 10											A was found. (AM: B→A)
Day 11	AM	D	-	-	-	-	-	-	-	-	Visit sequence counting
	PM	-	-	-	-	-	-	-	-	-	begins on Day 11, once
Day 12	AM	-	-	-	-	-	-	-	-	-	all four platforms were
	PM	-	-	-	-	-	-	-	-	-	visited and known.
Day 13	AM	D→	B	-	-	-	-	X	-	-	
	PM	-	-	-	-	-	-	-	-	-	
Day 14	AM	D→*	C*	D→ **	B→ **	A**	X	-	X**	-	*=1st focal monkey
	PM	-	-	-	-	-	-	-	-	-	**=2nd focal monkey
Day 15	AM	D→	A→	B	-	-	-	X	X	-	
	PM	B	-	-	-	-	-	-	-	-	
Day 16	AM	D→	B	-	-	-	-	X	-	-	
	PM	-	-	-	-	-	-	-	-	-	
Day 17	AM	B→	C→	A	-	-	-	X	X	-	
	PM	-	-	-	-	-	-	-	-	-	
Day 18	AM	B→	A	-	-	-	X	-	-	-	
	PM	B→	D	-	-	-	-	X	-	-	
Day 19	AM	-	-	-	-	-	-	-	-	-	
	PM	-	-	-	-	-	-	-	-	-	
Day 20	AM	B→	A	-	-	-	X	-	-	-	
	PM	-	-	-	-	-	-	-	-	-	
Day 21	AM	B→	A	-	-	-	X	-	-	-	
	PM	B	-	-	-	-	-	-	-	-	
Day 22	AM	D→	C→	D→	A	-	X	-	-	X	
	PM	-	-	-	-	-	-	-	-	-	
Day 23	AM	B→*	A*	B**	-	-	X	-	-	-	*=1st focal monkey
	PM	-	-	-	-	-	-	-	-	-	**=2nd focal monkey
Day 24	AM	-	-	-	-	-	-	-	-	-	
	PM	-	-	-	-	-	-	-	-	-	
Day 25	AM	B→	A→	D→	C	-	X	-	X	-	
	PM	-	-	-	-	-	-	-	-	-	
Day 26	AM	B→	D→	C	-	-	-	X	X	-	
	PM	-	-	-	-	-	-	X	-	-	
Day 27	AM	B→	D→	C	-	-	-	-	X	-	
	PM	B	-	-	-	-	-	-	-	-	

Table 6.5. Foraging pattern in Experiment 2: Different distances/Place constant/Higher *vs.* lower food reward.

Experimental days	Time of visit	1st visit	2nd visit	3rt visit (if any)	4th visit (if any)	5th visit (if any)	Short distance/ Less reward	Long distance/ More reward	Step-by-step movement	Back-tracking movement	Comments
Day 28	AM	-	-	-	-	-	-	-	-	-	
	PM	-	-	-	-	-	-	-	-	-	
Day 29	AM	B	-	-	-	-	-	-	-	-	
	PM	-	-	-	-	-	-	-	-	-	
Day 30	AM	-	-	-	-	-	-	-	-	-	
	PM	-	-	-	-	-	-	-	-	-	
Day 31	AM	B→	B	-	-	-	-	-	-	-	
	PM	B	-	-	-	-	-	-	-	-	
Day 32	AM	D→	C	-	-	-	X	-	-	-	
	PM	B	-	-	-	-	-	-	-	-	
Day 33	AM	D	-	-	-	-	-	-	-	-	
	PM	-	-	-	-	-	-	-	-	-	
Day 34	AM	D→	C	-	-	-	X	-	-	-	
	PM	-	-	-	-	-	-	-	-	-	
Day 35	AM	-	-	-	-	-	-	-	-	-	
	PM	-	-	-	-	-	-	-	-	-	
Day 36	AM	B→	A	-	-	-	X	-	-	-	
	PM	-	-	-	-	-	-	-	-	-	
Day 37	AM	-	-	-	-	-	-	-	-	-	
	PM	B	-	-	-	-	-	-	-	-	
Day 38	AM	A→	B	-	-	-	-	-	-	-	
	PM	-	-	-	-	-	-	-	-	-	
Day 39	AM	-	-	-	-	-	-	-	-	-	
	PM	-	-	-	-	-	-	-	-	-	
Day 40	AM	-	-	-	-	-	-	-	-	-	
	PM	-	-	-	-	-	-	-	-	-	
Day 41	AM	-	-	-	-	-	-	-	-	-	
	PM	-	-	-	-	-	-	-	-	-	
Day 42	AM	-	-	-	-	-	-	-	-	-	
	PM	B	-	-	-	-	-	-	-	-	
Day 43	AM	-	-	-	-	-	-	-	-	-	
	PM	-	-	-	-	-	-	-	-	-	
Total							10 (58.8%)	7 (41.2%)	6 (85.7%)	1 (14.3%)	= number of platform sequences (percentage).

Abbreviations: Each platform was designated with a capital letter from A to D; →= from platform "X" to the platform "Y;" AM= Ante meridian (from time of departure from sleeping site departure to 11:59); PM= Post meridian (from 12:00 to sleeping site arrival time).
Notes: a) Platforms B and D had more food rewards (eight bananas) than platforms A and C (fout bananas); b) The sequence of the 2ⁿᵈ focal monkey on Day 14 was discarded in the calculation due to the fact that the first platform visited might have not been recorded.

Table 6.5. continued.

Do white-faced capuchins travel between sequential feeding platforms using a distance-minimizing principle, and over the course of the field experiment do they reduce their circuity index in traveling between experimental platforms?

To explore if capuchins used a distance-minimizing principle when traveling between feeding platforms, I estimated the circuity indices to platforms that were visited in succession. In Experiment 2 (rectangular configuration, 80×180 m), the capuchins visited two platforms in succession 13 times, three platforms in succession five times, and four platforms in succession two times. The average circuity index was 1.39 at

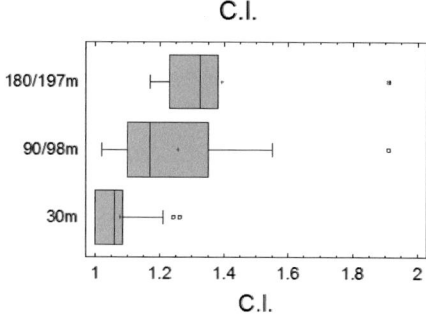

Figure 6.8. Comparison of the circuity indices of Experiment 2.

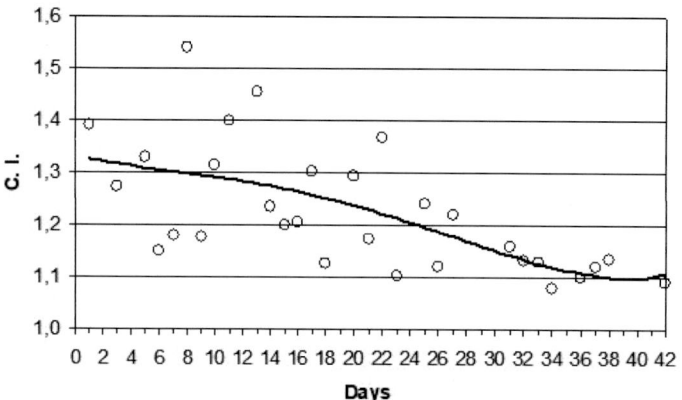

Figure 6.9. Learning curve with the circuity indices of Experiment 2 (*Note*: Fourth-degree polynomial line, $R^2=0.395$).

the largest distances between vertices (180 m and 197 m, *n*=6). The circuity index was 1.26 between platforms separated by 80 m (*n*=14). For the distances of 180/197 m, CI ranged from 1.10 and ≤1.25 on 60.7% of occasions. In the remaining 39.3% of cases, the CI was >1.25. At half the distance (90/98 m), the mean CI was 1.25. On only 5.5% of these occasions was the CI value <1.10 (1/18), whereas on 55.5% of occasions the CI ranged from 1.10 and ≤1.25 (10/18). Finally, at a distance of 30 m, the capuchins traveled only 8% more than the straight-line distance to their goal.

As in Experiment 1, I compared the CI values when the capuchins traveled different distances to reach the platforms in this rectangular configuration of Experiment 2. This was done in order to determine if CI values changed with decreasing distance to the target. I found that comparing distances of 180/197 m with distances of 90/98 m (*t*=1.27, d.f.=22, *p*>0.05) the differences in CI were not significant. However, a comparison of the CI values at 180/197 m *vs*. 30 m (*t*=4.73, d.f.=24, *p*<0.01), and 90/98 m *vs*. 30 m (*t*=3.35, d.f.=36, *p*<0.01) were showed significant, indicating that as in Experiment 1, capuchins used a more direct travel route as they approached the feeding sites. Fig 6.8 presents the pattern of CI variation according to the distances.

To understand how capuchins incorporate information over time, I also examined the evidence for learning based on the mean CI values over the 43 days of Experiment 2

(Fig. 6.9). During the first five days, the monkeys exhibited a CI of ~1.30. However, by Day 30 CI was reduced to a value between ~1.10 and ~1.20. A comparison of CI values during the first five days and the last five days of Experiment 2 indicates significant differences (t=2.96, d.f.=5, p<0.01). As in Experiment 1, these findings suggest that capuchins learned over time to adopt more efficient travel routes when moving between the experimental feeding platforms.

Do white-faced capuchins use traditional routes of travel to reach experimental feeding platforms or do they select novel travel routes?

To examine if capuchins formed new travel routes between experimental feeding sites or whether the monkeys continued to rely on habitual routes of travel to reach feeding platforms, I calculated the degree to which routes used by the capuchins during the natural field study also were used during Experiment 2 (see Urbani 2009: Appendix G). In this experiment, the area enclosed by the rectangular configuration (80 m x 180 m) was 1.4 ha. During this experiment, 2.3% of this area used by the capuchins was also used during the natural field study. In 29.3% of the area, the capuchins formed new travel routes during this experiment. Sixty-eight point four percent of the area was not used either during the natural field study or during this experiment.

A total of 36 quadrats (20 m x 20 m) were represented within the rectangular configuration of Experiment 2. In total, 36.1% of the quadrats (13/36) were crossed by the capuchins during this experiment, and 5.6% of the quadrats (2/36) included routes used during both the natural field study and Experiment 2. The remaining quadrats (58.3%; 21/36) contained neither novel routes nor routes used during the natural field study. The results of this experiment indicate that capuchins formed new routes of travel when moving between experimental platforms. However, these routes did not represent shortcuts and over the course of this experiment, the capuchins modified but continued to use the same travel segments between experimental platforms rather than develop more direct shortcuts. This supports the contention that capuchins formed a route-based spatial representation.

A total of seven areas used as nodes by the capuchins during the natural field study (March-August) were located in the 1.4 ha area within the perimeter of Experiment 2. These are nodes #3 and #4 used in March, Node #25 used in May, and nodes #41, #42, and #43 used in June (see Urbani 2009: Appendix D). Node #40 was crossed by a route that was not re-used during the experiment. The seven nodes were visited an average of 5.1±6.1 days during the course of the 43 days that lasted this experiment. A total of 0.9±0.9 nodes were visited per day. The seven nodes were crossed on 36 occasions (Node #3=1, Node #4=18, Node #25=3, Node #40=1, Node #41=8, Node #42=3, and Node #43=2; Table 6.6). These results are consistent with the contention that capuchins used a route-based system of spatial representation.

To examine the pattern of route use, I identified six route segments that were bounded by the nodes used in Experiment 2. The mean length of the route segments was 49.4±21.6 m. The mean distance of route segments ranged from 31.3 m on Day 18 to as high as 86.8 m on Day 25 (Table 6.7). Capuchins did not re-use route segments between these nodes during this experiment. Nonetheless, the route segments formed during this experiment bounded between nodes #40 and #42, and between nodes #41 and #42 were

Days:	1	3	5	6	7	8	9	11	14	15	16	17	18	20	23	25	26	27	29	31	34	36	37	42	Total	%
Node 3																						1			1	2.8
Node 4	1		1	1			1		1	1	1	1	1	1	1	1	1	1	1	1			1	1	18	50.0
Node 25		1				1		1																	3	8.3
Node 40																1									1	2.8
Node 41	1		1		1		1				1	1	1					1							8	22.2
Node 42					1		1									1									3	8.3
Node 43										1											1				2	5.6
	2	1	2	1	2	1	3	1	1	2	2	2	2	1	1	3	1	2	1	1	1	1	1	1	36	

Table 6.6. Nodes recorded during the natural field study used in Experiment 2.

	Node 4	Node 41	Node 42
Node 3	Day 15 (46.2 m)		
Node 4		Days 16, 18, 27 (62.4 m)	
Node 25		Day 18 (31.3 m)	Day 27 (37.5 m)
Node 40			Day 25 (86.8 m)
Node 41			Days 5, 9 (32.5 m)
Segment length average (m) 49.4 m			
SD	21.6 m		

Table 6.7. Route segments recorded during the natural field study used in Experiment 2.

used during the natural field study. In June, each of these route segments (between nodes #40-#42 and #41-#42) was used for three days. The route segment bounded between nodes #3 and #4 in this experiment also was previously used on two days during March. These data are consistent with a route-based spatial representation. These results also indicate that the monkeys formed and used novel routes of travel when traveling between the experimental platforms; however, the new route segments were not short-cuts between feeding platforms. In summary, as in Experiment 1, this is consistent with spatial information encoded as a route-based system.

Do white-faced capuchins select nearer platforms over more distant platforms? Do white-faced capuchins travel to more distant feeding platforms that contain higher food rewards preferentially over nearer feeding platforms that contain lower food rewards?

To test if distance was a primary factor in capuchin foraging decisions in Experiment 2, white-faced capuchins were presented with platforms that differed both in distance and food reward (Table 6.5). If distance is more salient in white-faced capuchins´ decision-making than food quantity, I expect the forager to preferentially visit a nearer platform containing a smaller food reward (80 m and four bananas) rather than visit a more distant platform containing a larger food reward (180/197 m and eight bananas). It was argued after Janson (1996, 2007) that wild capuchins tradeoff between distance and quantity information. In these previous studies, the capuchins selected sites with more food reward (min. four *vs.* max. 16 tangerines) (Janson 1996), and nearest feeding sites (range: 294 m to 665 m) independently of the amount of the food reward amount (Janson 2007). In a related study, Garber and Paciulli (1997) found that capuchins selected higher-rewarded feeding sites (three bananas) over lower-rewarded feeding sites (½ banana) when foraging within a food patch.

In Experiment 2, capuchins were found to select nearby platforms with lower food rewards over more distant platforms containing a greater food rewards. The capuchins also tended to select a higher reward platform as their first feeding platform, and then travel to a second platform located 80 m distant with a lower food reward (58.8% of the occasions; 10/17). In 41.2% (7/17) of the cases, the capuchins traveled from a first platform with a higher food reward to a second platform located at 180/197 m also with a higher food reward. However, these result are not significant (G_{adj}=0.52, d.f.=16, p>0.05). This suggests that depending on other factors such as satiation, either the size of the food reward (four bananas *vs.* eight bananas), or the distance to the platform (80 m *vs.* 180 m) could play an important role in determining which platform the capuchins visited second. Given that the first platform selected had a high food reward (eight bananas), it is possible that depending on the capuchins' previous behavior, they may have been satiated and thus not motivated to travel a greater distance to reach the more productive feeding site.

In terms of distance, the probability of selecting a closer feeding platform randomly once the first platform was visited was 33%. Capuchins selected the closer platform on 57.9% (11/19) of the occasions. The capuchins went to a more distant platform on 42.1% (8/19) of the occasions. After two platforms were visited the chance of randomly visiting a closer or a distant platform each was 50%. Capuchins opted for the closer platform on 71.4% (5/7) of occasions, and the most distant platform on 28.6% (2/7) of occasions. In terms of food quantity, the probability of selecting a feeding platform with a higher food reward randomly once the first platform was visited (and was a high-reward platform) was 33%. The capuchins choose the platform with a higher food reward in 36.8% (7/19) of the cases, and the platforms with the lower food reward in 63.2% (12/19) of the cases. Once two platforms with a higher food rewards or lower food rewards were visited the chance of randomly visiting platforms with a higher food reward or a lower food reward was 50% each. In these cases, capuchins chose platforms with higher food rewards at 71.4% (5/7) times and with lower food rewards in 28.6% (2/7) times.

When the variables distance and quantity are combined, the probability of choosing a feeding platform randomly once the first platform with a higher food reward (Platform B and Platform D) was visited is 33% for each of the following three categories: higher food reward/longer distance, lower food reward/shorter distance, and lower food reward/ longer distance. The capuchins went to the platforms with higher food reward/longer distance 33.3% of the time (6/18). The capuchins choose platforms with lower food reward/ shorter distance 55.6% of the time (10/18), and platforms with lower food reward/longer distance 11.1% of the time (2/18). After two platforms were visited, the chance of randomly visiting a platform of four different categories is 25%. These categories are platforms with higher food reward/longer distance, lower food reward/shorter distance, lower food reward/longer distance, and higher food reward/shorter distance. Capuchins opted for platforms with higher food reward/longer distance on 28.6% (2/7) of occasions, lower food reward/shorter distance on 14.3% (1/7) of occasions, lower food reward/longer distance on 14.3% (1/7) of occasions, and a higher food reward/shorter distance on 42.9% (3/7) of occasions. Given the small sample sizes, these results are not significant; however, in general, the capuchins preferentially selected nearer feeding sites regardless of whether they contained four or eight bananas.

Do white-faced capuchins exhibit a win-shift foraging rule when selecting experimental feeding platforms?

On those occasions when the capuchins visited the experimental platforms, I examined how capuchins visited them in succession (Table 6.5). By succession, I mean consecutively visiting three or more platforms. In 85.6% of the cases (6/7) capuchins selected an unvisited platform as their next platform, instead of re-visiting a recently-visited platform (Table 6.5). This is consistent with a win-shift foraging rule in which once a feeding site is exploited the forager travels to another feeding site of the same type. In only one of seven cases (14.3%) the capuchins backtrack to a previously visited platform. As in Experiment 1, once a platform was visited, the monkeys did not re-visit that platform during the test session.

Discussion

The main goal of this chapter was to examine the abilities of capuchins to relocate experimental feeding sites and to integrate spatial information in finding novel feeding sites. There currently exist relatively few natural or experimental field studies designed to evaluate spatial cognition in primates in large-scale space. The pioneering work of Janson and colleagues with wild black-horned capuchins (*Sapajus nigritus*) (Janson 1996, 1998, Janson and Di Bitetti 1997, Di Bitteti and Janson 2001) indicated that these monkeys integrated information on food quantity (min. four tangerines *vs.* max. 16 tangerines) to select distant feeding sites (>180 m apart). Further research found that capuchins' decisions can be viewed as a tradeoff between distance and food quantity, with capuchins tending to choose nearby feeding sites independently of the amount of their food reward (Janson 2007). In contrast, Garber and Paciulli (1997) and Garber (2000) indicated that in small-scale space (13 platforms at approximately 3.2 m apart), white-faced capuchins (*Cebus imitator*) quickly learned to select platforms that contained greater amounts of concealed food (½ banana *vs.* three bananas). In these experiments, the capuchins use a hierarchy of cues, they first learned absence *vs.* presence of food rewards, and then lower *vs.* higher amount of food rewards

In the present experimental field study, I found that capuchins integrated previously learned spatial information with novel spatial information to select new feeding sites. In Experiment 1, capuchins' likelihood of finding new feeding platforms was influenced by the spatial information they used during the natural field study. This is in accord with the contention that foragers search for new feeding sites in areas where previous foraging activity has been successful (Shettleworth 1998). According to Cheng and Spetch (1998), a forager relies on the use of fixed points in space to locate goals. These fixed points are landmarks or other physical features of the environment that serve as information beacons for navigation. These landmarks can be located near to sites and used for small-space navigation, or located distant to targets and used for navigation in large-scale space. In this context, the ability to relocate feeding sites is dependent on an association between single or multiple local landmark(s) and the goal. In some cases, a forager may use landmarks relationally to relocate their goal (Collet *et al.* 1986, Cheng and Sherry 1992, Newcombe and Huttenlocher 2000, Garber and Brown 2006). The results of my field experiments suggest that capuchins probably used landmarks learned prior to the placement of the feeding platforms or incorporated new spatial information over the course of days in order to locate the feeding platforms more efficiently.

In both experiments, I examined the pattern of sequential exploration of feeding platforms (Bateson and Kacelnik 1998). Decision-making implies a process in which a forager selects between two or more sets of information in order to select a behavioral response (Durkas 1998, Ydenberg 1998, Schuck-Paim and Kacelnik 2007). In this study, decision-making was examined by evaluating the movement patterns of the capuchins. In a study testing the ability of captive capuchins (*S. apella*) to form a search strategy once a feeding patch was depleted, the capuchins were found to avoid returning to previously exploited patches (De Lillo *et al.* 1997, 1998). In these studies, the authors suggested that capuchins represent each feeding patch as an individual goal. Once exploited, the monkeys move to the next available patch (De Lillo *et al.* 1997, 1998). The results of this monograph support the finding of De Lillo and colleagues. In Experiments 1 and 2 the capuchins visited the platforms using a win-shift foraging rule, and on only 10% of occasions revisited an experimental platform on the same day. The fact that white-faced capuchins explored unvisited platforms without returning to previously visited platforms indicates that the monkeys are able to plan and form an efficient search strategy when visiting multiple feeding patches.

In Experiments 1 and 2, once the first platform was visited, the capuchins selected the nearest platform most of the time. Moreover, in Experiment 2, this occurred despite the fact that the nearer platform had a lower food reward. These results are consistent with those recently found in a field experiment with black-horned capuchins (Janson 2007). In that study, platforms were located in a triangulated configuration (distances range: 294 m to 665 m; reward amount range: five to up to 80 pieces of tangerines or bananas). The black-horned capuchins selected the platforms that were located at a shorter distance independent of the amount of food reward (Janson 2007). Nevertheless, the findings of the field experiments of Janson (2007) and Urbani (2009, this study) are limited due to the possibility that based on their earlier feeding behavior the capuchins were satiated and not motivated to travel between platforms, particularly to those located at longer distances. However, the general trend is for capuchins to preferentially select nearer feeding platforms regardless of the quantity of food reward. Additional research testing the tradeoff between food quantity and distance is a research priority.

In both field experiments, the capuchins used nodes and route segments recorded during the natural field study to locate experimental feeding platforms. Although there also was evidence that the capuchins used new routes to reach experimental platforms. However, they did not use shortcuts or straight-line route segments when traveling between the feeding platforms (CI=1.27). This result is consistent with a route-based spatial representation in large-scale space.

The monkeys reduced their circuity index over time to reach previously visited platforms. On average CI was ~1.30, but decreased to ~1.10/~1.20 during the final five days of each field experiment. In addition, with decreasing travel distance to a reward platform (Exp. 1: 60/85 m, and Exp 2: 90/98 m), the monkeys traveled in a more distance-minimizing trajectory, reducing CI from 1.24 (Exp. 1) and 1.39 (Exp.2) to 1.09 (Exp. 1) and 1.25 (Exp. 2) at half the distance. When located 30 m from the target, the monkeys only deviated from the most direct route by 5%. These results indicate that in large-scale space the capuchins are using a route-based spatial representation, and in small-scale space, they are using a coordinate-based spatial representation.

In sum, the results of this chapter indicate that the capuchins used previously learned spatial information to locate new feeding sites. In their decision-making, the distance was more salient than food quantity in selecting the second, third or fourth feeding site. Thus, in general, capuchins choose feeding sites located at a shorter distance (Exp. 1: 120 m *vs.* 169 m; Exp. 2: 80 m *vs.* 180/197 m) independently of the amount of food (four or eight bananas) at these sites. Capuchins formed a foraging strategy in which once the first feeding site was visited, the next feeding site was another experimental platform. This is consistent with a win-shift foraging rule and traplining. There also was evidence of a learning curve. That is, the capuchins reduced the circuity index over time to return to a given feeding platform. Thus, in large-scale space, capuchins relied on a route-based spatial representation. In small-scale space, as stated above, there is evidence, such as the use of direct travel routes once the monkeys were within 30 m of the feeding/resting tree, to suggest that capuchins encode spatial information in the form of a coordinate-based spatial representation. This pattern of spatial representation appears to support the model proposed by Poucet (1993). However, there also are data that do not fully support the ability of white-faced capuchins to form a coordinate-based map. This includes the inability of capuchins to take novel short-cuts (high circuity indices) for reaching the platforms, the use of nodes as switch points to re-orient travel, the fact that throughout the study period the capuchins were not able to form distance-efficient itineraries, and that possibly the monkeys used local landmarks rather than a coordinate-based spatial representation to re-locate feeding platforms.

NAVIGATING WITH WHITE-FACED CAPUCHIN MONKEYS

7

Conclusions

The main objective of this research was to address a set of anthropological questions concerning feeding ecology, spatial memory, and the evolution of cognitive abilities in non-human primates. I examined spatial mapping abilities in wild Costa Rican capuchin monkeys. There exists a growing body of evidence that non-human primates integrate social and ecological information in decision-making (Garber *et al.* 2009). Decision-making in non-human primates is hierarchical; that is, in different situations certain information may be more salient than other information in deciding which feeding site to exploit or whether to co-feed with a particular group member. In this study, I have examined primate decision-making using two approaches. I integrated data collected during an 8-month natural field study of capuchin spatial mapping with two field experiments that tested how capuchins integrate spatial information and quantity information in decision-making.

It has been suggested that capuchin monkeys represent an important animal model for understanding issues of cognitive abilities, decision-making, and encephalization in primates (Visalberghi 1990; Fragaszy *et al.* 2004). Capuchins exhibit a set of behavioral and cognitive traits rarely found in non-ape primates (Visalberghi and McGrew 1997). Compared to other New World primates, capuchins exhibit enhanced manual dexterity and the capacity to use and manipulate tools in both wild and captive settings (Costello and Fragaszy 1998; Boinski *et al.* 2000, Spinozzi *et al.* 2004, Garber and Brown 2004, Moura and Lee 2004, Fragaszy *et al.* 2004, Ottoni and Mannu 2008, Rodrigues-Canale *et al.* 2009). As indicated by Hershkovitz (1977), capuchins are distinctive among Neotropical primates due to their complex expression of cerebral fissures and large brain size. Capuchins present a high ratio of the brain relative to body size (Jerison 1973). Capuchins along with chimpanzees engage in extractive foraging, hunting (Rose 1997), and tool use, a set of behavioral patterns that are analogous to hominins (see Ottoni and Izar 2008). Capuchins are reported to display local behavioral "traditions" (Fedigan 1990, Panger *et al.* 2002, Perry *et al.* 2003, Perry *et al.* 2004) including "games," hand-sniffing, and sucking body parts. These local traditions are thought to be transmitted through a process of social learning (Perry *et al.* 2003). Consequently, capuchins have been the focus of numerous captive, natural, and experimental field studies designed to explore questions of primate cognition (e.g. Visalberghi, 1990, Janson 1996, 1998; Janson and Di Bitetti 1997, 2001; De Lillo *et al.* 1998, Urbani 1999, Potì 2000, Ottoni and Mannu 2001, Di Bitetti and Janson 2001, Garber and Paciulli 1997, Garber and Brown 2006; Moura and Lee 2004; Fragaszy *et al.* 2004).

Several authors have argued that travel in non-human primates is goal-directed, with individuals using habitual routes of travel within their home ranges. It remains unclear, however, whether primates that exploit large home ranges and travel to distant feeding sites acquire and use spatial information differently than primates that exploit smaller home ranges. In addition, little is known concerning the degree to which individuals encode spatial information using a route-based or a coordinate-based spatial representation. Several primates, including prosimians, New World monkeys, Old World monkeys, and apes, have been described as using route-based navigation (e.g. Milton 1980, Garber 1989, 2000, Janson 1996, 1998, Di Fiori and Suárez 2007, Noser and Byrne 2007). Route-based navigation is defined as a spatial representation in which a forager is expected to encode and integrate multiple route segments by orienting to nodes or areas of the forest that contain natural landmarks. In contrast, using a coordinate-based spatial representation, a forager is expected to encode information as true distances and true bearings in order to compute shortcuts to reach out-of-view targets. It has been suggested that tamarins and capuchins plausibly use a coordinate-based spatial representation in small-scale space and that chimpanzees use a coordinate-based spatial representation in large-scale space (Garber 1989, 2000, Garber and Brown 2006, Boesch and Boesch 1984, Normand and Boesch 2009, Urbani 2009, this study).

To test how foragers spatially represent their environment, Poucet (1993) developed a model of navigation. Poucet (1993) hypothesized that when traveling in small-scale space, primate foragers have the opportunity to encode spatial information within a coordinate-based system, whereas when traveling in large-scale space they are more likely to use a route-based spatial representation. He argues that a critical factor in the ability to construct a coordinate-based spatial representation is the opportunity to obtain views of the same set of landmarks and goals from multiple directions. In tropical forests, such views may be restricted by the denseness of the canopy whereas, in more open habitats, unobstructed views may be available for distances of hundreds of meters (Garber 2000, Dominy *et al.* 2001). A forager using a coordinate-based spatial representation is expected to encode different views of particular targets and landmarks, constructing a "view from above," and use this information to compute novel short-cut routes of travel to reach out-of-view feeding sites. Poucet (1993) argues, however, that when navigating in large-scale space animals are unable to obtain views of goals and landmarks from multiple locations and therefore are expected to employ a route-based spatial representation in which individuals orient to and use and re-use a limited set of familiar routes of travel and nodes to reach major feeding/resting sites. These nodes contain specific features of the environment (i.e. landmarks) that function as navigation switch points to connect travel routes. By associating the spatial position of a common set of landmarks relative to important targets, the forager can use these cues to re-orient travel. A route-based representation restricts the forager to a set of "known" travel routes and nodes in going from one target or goal to another. As stated above, a major question in the study of primate cognitive ecology is the degree to which different primate species form route-based maps and/or coordinate-based maps (Poucet 1993, Garber 2000). In this project, I test Poucet's model of spatial representations in a wild population of white-faced capuchins.

In this research, each month approximately three species accounted for ≥50% of capuchin plant feeding time. I defined this subset of tree species or highly selected foods (HSFs). In addition, feeding and foraging on animal matter accounted for approximately 43% of the capuchin monthly activity budget. This pattern of feeding and foraging was stable throughout the year. The exploitation of highly selected foods involves cognitive abilities associated with spatial memory to recall the location of feeding sites. Data presented in this study indicate that the capuchins were able to relocate important feeding/resting trees within their home range by using a small set of travel route segments (mean=12.4±.8.5) each month.

Several datasets collected in this study support the contention that white-faced capuchins used a route-based map when traveling in large-scale space with no visibility between goals (111±81 m). During the natural field study, the capuchins tended to select the nearest neighbor trees of a set of target species. The ability to find the nearest neighbor trees requires that a forager has knowledge of the relative location of individual trees fruiting at the same time. This foraging pattern reduces random search. Moreover, when traveling between feeding/resting sites the capuchins did not take novel shortcuts or the most direct travel routes. I found that in large-scale space, capuchins relied on a small set of nodes to orient travel along route segments. For example, the white-faced capuchins used a total of 53 nodes during the natural field study, crossing six of these nodes each day. Twenty-one of these nodes were visited for a minimum of 10 days.

Route segments used by the capuchins averaged 90 m in length. On average, route segments were re-used 3.9 times per month. The number of individual route segments and nodes varied from month to month. However, as the study progress, the re-use of nodes increased significantly. This is consistent with the assumption that the monkeys relied on landmarks or other topological features of the environment to reach their targets. I also found that the routes taken by the capuchins were on average 42% greater than the straight-line distance (CI=1.42). The travel route taken by the capuchins between feeding/resting trees and the distance traveled is affected by foraging for insects. Taken together, these data are consistent with the hypothesis that white-faced capuchins employ a route-based spatial representation in large-scale space.

When capuchins approached the vicinity of feeding/resting trees (30 m), they took highly direct routes to reach them (CI index at 30 m=1.06) before reaching the perimeters of those trees (mean=12.6 m). In addition, the capuchins visited major feeding/resting trees from multiple directions. For example, on average the major feeding/resting trees were visited on 14 on different occasions, 10 of which were from different cardinal directions. These data suggest that the capuchins form a mental representation of major feeding/resting trees and as they approached these sites form direct routes at the vicinity of these targets. In small-scale space, this is consistent with a coordinate-based spatial representation.

During three months of this research, I conducted two experimental studies designed to test the capuchins´ abilities to relocate experimental feeding platforms and to integrate spatial information and food quantity in decision-making. The results of the field experiments indicate that white-faced capuchins used spatial information encoded during the natural field study to encounter the experimental feeding sites. For example, the order in which the platforms were discovered in Experiment 1 correlated with the location of route segments and nodes used during the natural field study. In the field experiments, the capuchins used nodes and route segments traveled during the natural field study to reach experimental platforms and did not use novel shortcuts or the most direct travel routes between platforms. In field experiments, in general, the capuchins selected nearer feeding platforms regardless of the quantity of the food reward.

Another finding of this research was evidence that the capuchins exhibited a learning curve in developing more efficient travel routes when navigating to the experimental platforms. Over time, the capuchins modified their travel route to decrease the circuity index (CI). For example, in Experiment 2, CI went from ~1.30 in the first five days to ~1.10 for the last five days of the experiment. Modification of travel routes, however, did not involve the use of novel shortcuts. These data support the contention that the capuchins encoded information on the location of feeding platforms and the distance between feeding platforms, and used this information in selecting feeding sites. The behavior of the capuchins in the field experiments was consistent with a route-based spatial representation. As in the natural field study, as capuchins move closer to the feeding platforms (30 m), travel became more straight-line and direct.

In large-scale space, experimental field studies of *S. nigritus* indicate that they use spatial information and quantity information in decision-making (Janson 1996, 1998, 2007, Janson and Di Bitetti 1997, Di Bitetti and Janson 2001). Similarly, based on an experimental field study in which wild white-faced capuchins (*C. imitator*) were required to use a set of landmark cues (2 m pink/yellow poles) to predict the location of hidden food rewards located on two of eight feeding platforms, Garber and Brown (2006) found that foragers used a combination of two and/or three landmark cues to compute the location of the reward platforms. These authors argued that using different views of the same set of landmarks, the capuchins were able to mentally rotate landmarks in small-scale space.

Chimpanzees (Boesch and Boesch 1984, Normand and Boesch 2009), tamarins (Garber 1989, 2000), and capuchins (Urbani 2009, this study) had been argued to use both route-based and coordinate-based maps at different spatial scales. However, it is likely that such a similar set of representation are used by other primate genera as well (Tomasello and Call 1997, Garber 2000, Byrne 2000, Janson and DiBitetti 1997). *Homo* is another primate genus that uses a combination of route-based and coordinate-based navigation.

Milton (2000) suggested that Amazonian hunters use particular features of the forest like ridges, creeks, and rivers as landmarks or trails. These physical features of the environment serve as reference points, nodes, and route segments that are used as reference points to navigate though the forest. In contrast, Walbiri aborigines have been reported to use precise notions of distance and bearing in order to move within an open environment such as the Simpson Desert of north-central Australia (Lewis 1976). Lewis (1976) reported that based on 34 experiments in which five individuals tested their navigation skills in open large space, they showed an average of 13.7° of error in reaching places located at distances of approximately 200–300 km (min: 10 km, max: 670 km). As indicated by McCabe (2000) the processes of deciding where to move among the Ngisonyoka, the Turkana nomads of East Africa, are influenced by the location of foraging areas, water ponds, areas perceived as "prophetic" and the consideration of security issues. The Turkana people travel between these areas depending on seasonality. They conceive their territory as dry season areas, wet season areas, and drought reserve area.

In the case of early humans and australopithecines, the shift to a broader diet composed of tubers, fruits, seeds, and meat, and travel within open and forested environments to feeding places and sites with lithic sources is suggested to have resulted in the expansion of home range area and arguably larger body size (Leonard and Robertson 2000, Ambrose 2001, Potts 2004). The possible differences in the use of home ranges by hominins may reflect changes in cognitive abilities (Mann 1981, Mitchen 2003, Boehm 2004; see Leigh 2004 for links between brain evolution and cognition). Changes in the hierarchical use of information or the ability to integrate different sets of information between *Australopithecus* and *Homo* help to explain potential variability in foraging strategies and use of space among these taxa, (McCabe 2000, Leonard and Robertson 2000, Ambrose 2001, Potts 2004). The study of spatial cognition in non-human primates is a useful framework for understanding the evolution of cognition and decision-making in human ancestors

In Chapter 1, I outlined a set of hypotheses associated with spatial memory. Based on this research (Table 7.1), I conclude the following, white-faced capuchins travel to feeding/ resting sites using more direct routes in small-scale space. Near the feeding/resting sites, their travel itineraries show low circuity indices. This is consistent, in part, with the use of coordinate-based spatial representation at this scale. In large-scale space, capuchins rely on longer route segments and nodes and have high circuity indices when traveling between feeding/resting trees. This supports the use of a route-based spatial representation.

This monograph offers a set of new insights and research questions concerning spatial mapping in primates. In terms of natural field studies, primatologists need to expand the database of problem-oriented field research on primate spatial memory. Data should include not only information on navigation skills for a range of non-human primate taxa but also detailed information on the ontogeny of spatial mapping. For example, in humans, Liben (1981, 1988) indicates that the storage and integration of spatial information reach its maximum competence during adolescence/adulthood after other cognitive skills, especially linguistic proficiency is fully acquired. Liben and Downs (1989) found, however, that map conceptualization, identification, and utilization appear relatively early in childhood (<5 years old) but proficiency increased in accordance with lifespan.

Other important research questions for continued study involve intra-specific variability in spatial memory and navigation patterns in primate populations living in different habitats. For example, white-faced capuchins living in seasonally dry environments such as Santa Rosa National Park, Costa Rica may face different challenges (such as marked dry season and capuchins ranging tied with limited water ponds) than capuchins living in rainforests. Such data will provide information on behavioral flexibility in patterns of habitat utilization and spatial memory. In addition, the examination of how primates encode new spatial information, such as in our experimental field studies, will provide important insight into areas when confronting changes in their home ranges due to natural or anthropogenic disturbance. In a study, Janmaat *et al* (2008) found that when sooty mangabeys (*Lophocebus albigena*) shifted to a new foraging area, the monkeys increased daily path length and were less efficient in locating the most common feeding trees such as *Ficus sansibarica*. Months later, the mangabeys increased their knowledge of the area and reduced daily path length. Qualitatively, the authors suggest that the change occurred because the monkeys were able to integrate new spatial information in decision-making.

Methodologically, a strength of this research project involved the construction of detailed maps from which to analyze capuchin spatial behavior. Considering that standard

GPS devices may have limited precision (≥20 m) (Urbani and P. A. Garber Pers. obs.) under dense forest canopies, the use of *in situ* 10 m x 10 m grid systems based on accurate and detailed mapping using flagging marks every 10 m with Y-X coordinates in the forest, allowed me to record primate locations with higher precision. In the future, the use of new automated radio telemetry systems may represent the best alternative for a detailed study of primate movement patterns and spatial memory (e.g. Crofoot *et al.* 2008). In addition, the use of detailed cartographic information and satellite images when available may also serve to help identify features of the environment used as landmarks for navigation. For example, Suárez (2003) found that the habitual routes used by spider monkeys (*Ateles belzebuth*) coincided with the ridges of hills located within their home ranges. MacKinnon (1974) similarly argued that orangutans (*Pongo pygmaeus*) used streams and ridges for navigation.

Field studies of primate spatial memory should include experimental approaches. This is due to the fact that field experiments allow the researcher to control ecological information presented to the foragers. Field experiments can directly test the manner in which temporal and spatial information are used hierarchically in decision-making. This includes experimental studies designed to determine whether individuals select platforms with a consistent food reward over platforms with a higher but less predictable food reward. Also, experiments testing the effects of age, sex, and rank on decision-making are critical for understanding how social factors affect exploration and spatial navigation.

In this research, I integrated natural field observations and field experiments to examine the ability of a New World primate forager to develop spatial representations in large and small-scale space (Table 7.1). I tested if capuchins used a combination of route-based and coordinate-based spatial representations. This depends on the ability of the monkeys to encode landmark cues and physical elements of the space within their home range. This monograph provides empirical evidence that capuchins navigational skills are most consistent with Poucet's (1993) model of the use of a route-based spatial representation in large-scale space and a coordinate-based spatial representation in small-scale space.

However, the evidence is also consistent with using route-based spatial representations in large-scale and small-scale spaces. For instance, based on data during the field experiments, the capuchins failed to adjust their travel path to take more direct and novel routes between feeding sites. Moreover, during the natural field study, they did not travel using distance-minimizing itineraries when moving between feeding/resting sites or travel directly to feeding sites from distances of greater than 55±5m. It is also possible that the use of longer than straight-line travel routes enable capuchins to collect ecological information about food availability, monitor the fruiting status of trees, and obtain information on insect distribution. Furthermore, among non-human primates, including white-faced capuchins, the energetic cost of traveling an additional 100 m, 200 m, or even 500 m seems to be relatively low (Kowalewski *et al.* 2009). Accordingly, it may be more efficient for foragers to take a variety of relatively efficient travel routes rather than direct shortcuts between feeding/resting sites if by doing so they obtain important resource information. In addition, in an experimental field study of spatial cognition in this same group of white-faced capuchins, Garber and Brown (2006) found the ability of the capuchins to use the relative position of local landmarks as "places" to calculate

Hypotheses (H)	Outcome	
H_1: If capuchins use a coordinate-based spatial representation, then they are expected to	$H_{1.1}$) travel to trees in a relatively straight-line fashion, using shortcuts or direct novel routes of travel.	X (In small-scale space only)
H_2: If capuchins use a route-based spatial representation, then they are expected to	$H_{2.1}$) travel to previously visited trees by re-using a set of travel route segments.	X
	$H_{2.2}$) utilize particular forest places of the forest such nodes as switch points to re-orient travel.	X
H_3: During Experiments 1 and 2, if capuchins primarily make foraging decisions based on distance rather than the amount of food reward, then	- When the quantity of the food reward is identical on platforms located at similar distances then, $H_{3.1}$) the capuchins are expected to visit nearer feeding sites once a previous feeding platform was visited.	X
	- At different distances and different amounts of food reward, it is expected that capuchins, $H_{3.2}$) visit the nearer feeding site even when this involves expectations of lower food reward.	X (Limited dataset)
H_4: – If capuchins use a route-based spatial representation, then during field experiments: $H_{4.1}$ – If capuchins use a coordinate-based spatial representation, then during the same field experiments: $H_{4.2}$	$H_{4.1}$) -it is expected that they will continue to re-orient travel at commonly used nodes to reach feeding platforms even when these routes are characterized by increased travel distance. – it is expected to have higher circuity indices while traveling to feeding platforms.	X
	$H_{4.2}$) -they are expected to compute novel short-cuts and more direct routes between experimental feeding sites. – it is expected to have lower circuity indices while traveling to feeding platforms.	X (In small-scale space only)

Note: X=found

Table 7.1. Evaluation of the hypotheses tested in this study (see Chapter 1).

direction and distances to their goals in small-scale space (an area with a diameter of 8 m). Thus, in small-scale space, capuchins may use a single or a set of visual cue(s) to reach their targets (Garber and Brown 2006). This would require individuals to encode large numbers of near-to-site landmarks. If this is correct, then Poucet´s model may be too mechanistic to explain variability in the use of different types of spatial representations at different spatial scales.

References

Abreu F, Garber PA, Souto A, Presotto A, Schiel N. (2021). Navigating in a challenging semiarid environment: the use of a route-based mental map by a small-bodied neotropical primate. *Animal Cognition* 24: 629–643.

Adcock J (2001). *Animal Behavior: An Evolutionary Approach*. Sinauer Associates.

Agoramoorthy G, Hsu, M.J (1995). Population status and conservation of red howling monkeys and white-fronted capuchin monkeys in Trinidad. *Folia Primatologica* 64: 158–162.

Aiello LC, Wheeler P (1995). The expensive-tissue hypothesis: The brain and the digestive system in human and primate evolution. *Current Anthropology* 36: 199–221.

Albuja L, Arcos R (2007). Evaluación de las poblaciones de *Cebus albifrons* cf. *aequatorialis* en los bosques suroccidentales Ecuatorianos. *Politécnica Biología* 7: 58–67.

Allman JM (2009). *Evolving brains*. New York: Scientific American Library.

Altman, J. (1997). *Development of the cerebellar system in relation to its evolution, structure, and functions*. Boca Raton: CRC Press, 783 pp.

Altmann J (1974). Observational study of behaviour: Sampling methods. *Behaviour* 49: 227–265.

Altmann SA (1974). Baboons, space, time, and energy. *Amer. Zool.* 14: 221–248.

Altmann SA, Altmann J (1970). *Baboon ecology. African field research*. Chicago: University of Chicago Press.

Ambrose SH (2001). Paleolithic technology and human evolution. *Science* 291: 1748–1753.

Aquino R, Bodner R. E., Pezo E.(2000). Evaluación de primates en la cuenca del río Pucacuro, Amazonía peruana. *In:* San Martín-Howard, F., García-Podestá, M. (eds.), *La primatología en el Perú, Vol. II*. Universidad Mayor de San Marcos, pp. 102–110.

Aquino R, Cornejo FM, Heymann EW (2013) Primate abundance and habitat preferences on the lower Urubamba and Tambo rivers, central-eastern Peruvian Amazonia. *Primates* 54: 377–383.

Aquino R. (1990). La fauna primatológica en áreas de Jenaro Herrera. *In:* Castro-Rodríguez, N. E. (eds.), *La primatología en el Perú, Vol. I*. Universidad Mayor de San Marcos, pp. 122–134.

Baerends GP (1941). Fortpflanzungsverhalten und Orientierung der Grabwespe *Ammophila campestris. Tijdschrift voor entomologie* 84: 68–275.

Baker M (1998). *Fur rubbing as evidence for medical plat use by capuchin monkeys (*Cebus capucinus*): Ecological, social and cognitive aspects of the behavior.* Riverside: University of California, PhD dissertation.

Baldwin JD, Baldwin JI (1976). Primate populations in Chiriqui, Panama. *In:* Thorington RW Jr., Heltne PG (eds.), *Neotropical Primates: Field Studies and Conservation.* Washington, DC: National Academy of Sciences, pp. 20–31.

Barlow, J. (2002). *The cerebellum and adaptive control.* Cambridge and New York: Cambridge University Press, 340 pp.

Barns A. (1923). *Across the Great Craterland to the Congo.* London: Ernest Benn, 276 p.

Barton RA (2009). Primate brain evolution: Cognitive demands of foraging or of social life?. *In:* Garber PA, Boinski S (eds.), *On the move: How and why animals travel in groups.* Chicago: Univ Chicago Press, pp. 491–518, 204–237.

Bates L, Byrne R (2004). How do chimpanzees (*Pan troglodytes*) find the resources they need in the African moist forest? *Folia Primatologica* 75 (Suppl. 1): 96.

Benhamou S (1996) No evidence for cognitive mapping in rats. *Animal Behavior* 52: 201–212.

Benhamou S, Sauve J, Bovet P (1990). Spatial memory in large-scale movements: Efficiency and limitations of the egocentric coding process. *Journal Theoretical Biology* 145: 1–12.

Bennet ATD. (1996). Do animals have cognitive maps. *Journal of Experimental Biology* 199: 219–224.

Bergeson DJ (1996). The positional behavior and support use of three Costa Rican primates. *IPS/ASP Congress* Abstracts, p. 533.

Bernstein IS (1964). A field study of the activities of howler monkey. *Animal Behaviour* 12(1): 92–97.

Betancourt D, Torres F, Fernández J. (1999) La sarrapia: importancia y utilización. *FONAIAP Divulga,* 64: http://www.ceniap.gov.ve/publica/divulga/fd64/texto/sarrapia.htm (Consulted December 2008).

Bezanson M (2005). Ontogenetic patterns of positional behavior in *Cebus capucinus* and *Alouatta palliata. American Journal of Physical Anthropology* (Suppl 40): 74.

Bicca-Marques JC (1999). *Cognitive aspects of within-patch foraging decisions in wild diurnal and nocturnal New World primates.* University of Illinois at Urbana-Champaign, PhD dissertation.

Bicca-Marques JC, Garber PA (2003). Experimental field study of the relative costs and benefits to wild tamarins (*Saguinus imperator* and *S. fuscicollis*) of exploiting contestable food patches as single- and mixed-species troops. *American Journal of Primatology* 60: 139–153.

Bicca-Marques JC, Garber PA (2004). Use of spatial, visual, and olfactory information during foraging in wild nocturnal and diurnal anthropoids: A field experiment comparing *Aotus, Callicebus,* and *Saguinus. American Journal of Primatology* 62: 171–187.

Bicca-Marques JC, Garber PA (2005). Use of social and ecological information in tamarin foraging decisions. *International Journal of Primatology* 26: 1321–1344.

Boehm C (2004). Large game-hunting and the evolution of human sociality. *In:* Sussman RW, Chapman AR (eds.), *The origins and nature of sociality.* New York: Aldine de Gruyter, pp. 270–287.

Boesch C (2005). Joint cooperative hunting among wild chimpanzees: taking natural observations seriously. *Behavioral and Brain Sciences* 28(5): 692–693.

Boesch C, Boesch H (1984). Mental map in wild chimpanzees: An analysis of hammer transports for nut cracking. *Primates* 25: 160–170.

Boinski S, Garber PA (eds.) (2000). *On The Move: How and Why Animals Travel in Groups.* Chicago: Univ Chicago Press, xi + 811 p.

Boinski S, Kauffman L, Ehmke E, Schet S, Vreedzaam A (2005). Dispersal patterns among three species of squirrel monkeys (*Saimiri oerstedii, S. boliviensis* and *S. sciureus*): I. Divergent costs and benefits. *Behaviour* 142: 525–632.

Boinski S, Quatrone RQ, Swartz H (2009). Substrate and tool use by brown capuchins in Suriname: Ecological contexts and cognitive bases. *American Anthropologist* 102: 741–761.

Bolaños R, Watson V, Tosi J (1993). *Mapa Ecológico de Costa Rica según el Sitema de Clasificación de Zonas de Vida del Mundo.* Centro Científico Tropical-ICE, San José, Costa Rica. Scale 1: 200.000.

Bolt, LM., Brandt LE, Molina RL, Schreier AM. (2021). Maderas Rainforest Conservancy: A One Health approach to conservation. *American Journal of Primatology* 84: e23293.

Bonaccorso FJ, Glanz WE, Sandford CM. (1980). Feeding assemblages of mammals at fruiting *Dipteryx panamensis* (Papilionaceae) trees in Panama: seed predation, dispersal, and parasitism. *Revista de Biologia Tropical* 28: 61–72.

Boyer D, Miramontes O, Ramos-Fernández G, Mateos JL, Cocho G (2004). Modeling the searching behavior of social monkeys. *Physica A: Statistical Mechanics and its Applications* 342: 329–335.

Braithwaite VA (1998). Spatial memory, landmark use and orientation in fish. In: Healy SD (ed.), *Spatial Representation in Animals.* Oxford: Oxford University Press, pp. Brown AD, Chalukian SC, Malmierca LM, Calillas OJ (1986). Habitat structure and feeding behavior of Cebus apella (Cebidae) in El Rey National Park, Argentina. *In:* Taub DM, King FA (eds.), *Current Perspectives in Primate Social Dynamics.* New York: Van Nostrand Reinhold Co, pp 137–151.

Brockman DK, van Schaik CP (2005). *Seasonality in primates: Studies of living and extinct human and non-human primates.* New York: Cambridge Univiversity Press. 590 pp.

Buckley JS (1983). The feeding behavior, social behavior and ecology of the white-faced monkey, *Cebus capucinus*, at Trujillo, Northern Honduras, Central America. *Dissertation Abstracts International* A44(4): 1143.

Burt WH (1940). Territorial behavior and populations of some small mammals in Southern Michigan. *Miscellaneous Publications of the Museum of Zoology, University of Michigan* 45: 1–58.

Burt WH (1943). Territoriality and home range concepts as applied to mammals. *Journal of Mammalogy* 24: 346–352.

Byrne RW (1979). Memory for urban geography. *Quarterly Journal Experimental Psychology* 39: 147–154.

Byrne RW (1982). Geographical knowledge and orientation. *In:* Ellis A (ed.), *Normality and pathology of cognitive function.* London: Academic Press, pp. 239–264.

Byrne RW (2009). How monkeys find their way: Leadership, coordination, and cognitive maps of African baboons. *In:* Garber PA, Boinski S (eds.), *On the move: How and why animals travel in groups.* Chicago: University Chicago Press, pp. 491–518.

Byrne RW, Whiten A (1988). Note on Milton (1988). *In:* Byrne RW, Whiten A (eds.), *Machiavellian intelligence: Social expertise and the evolution of intellect in monkeys, apes, and humans.* Oxford: Clarendon Press, pp. 285.

Byrne RW, Whiten A (eds.) (1988). *Machiavellian Intelligence: Social Expertise and the Evolution of Intellect in Monkeys, Apes, and Humans.* Oxford: Clarendon Press, xv, 413 p.

Call J (2005). The self and other: a missing link in comparative social cognition. *In:* Terrace HS, Metcalfe J (eds.), *The Missing Link in Cognition: Origins of Self-Reflective Consciousness.* New York: Oxford Univ Press, pp. 321–341.

Call J (2009). Representing space and object in monkeys and apes. *Cognitive Science* 24: 397–442.

Cameron RC, Wiltshire C, Foley N Dougherty X, Aramayo, Rea L (1989). Goeldi's Monkey and other primates in northern Bolivia. *Primate Conservation* 10: 62–70.

Carpenter CR (1934). A Field Study of the Behavior and Social relations of Howling Monkeys. *Comparative Psychology Monographs* (The John Hopkins Press) 10(2): 1–168.

Chapman CA (1987). Flexibility in diets of three species of Costa Rican primates. *Folia Primatologica* 49: 90–105.

Chapman CA (1988). Patterns of foraging and range use by three species of Neotropical primates. *Primates* 29: 177–194.

Chapman CA, Chapman LJ, McLaughlan RL (1989). Multiple central place foraging by spider monkeys: Travel consequences of using many sleeping sites. *Oecologia* 79: 506–511.

Chapman CA, Fedigan LM (1990. Dietary differences between neighboring Cebus capucinus groups: Local traditions, food availability or responses to food profitability? *Folia Primatologica* 54: 177–186.

Cheng K, Sherry D. (1992). Landmark-based spatial memory in birds (*Parus atricapillus* and *Columba livia*): The use of edges and distances to represent spatial positions. *Journal of Comparative Psychology,* 106: 331–341.

Cheng K, Spetch M (1998). Mechanisms of landmark use in mammals and birds. *In:* Healy S (ed.): *Spatial prepresentation in animals.* Oxford: Oxford University Press, pp. 1–17.

Cheng K. (1986). A purely geometric module in the rat´s spatial representation. *Cognition,* 23: 149–178.

Chun SLM. (2008). *The Utility of Digital Aerial Surveys in Censusing* Dipteryx panamensis, *the Key Food and Nesting Tree of the Endangered Great Green Macaw (*Ara ambigua*) in Costa Rica.* Duke University, PhD dissertation, 163 p.

Clutton-Brock TH, Harvey PH (1977). Primate ecology and social organization. *Journal of Zoology* (London) 183:1–39.

Collet TS, Cartwright BA, Smith BA. (1986). Landmark learning and visuo-spatial memories in gerbils. *Journal of Comparative Physiology* A158: 835–851.

Collet TS, Zeil J (1998). Places and landmarks: an arthropod perspective. *In:* Healy S (ed.), *Spatial prepresentation in animals.* Oxford: Oxford University Press, pp. 18–53.

Conklin NL, Wrangham RW (1994). The value of figs to a hind-gut fermenting frugivore: A nutritional analysis. *Biochemical Systematics and Ecology* 22: 137–151.

Costello M, Fragaszy D (1988). Prehension in *Cebus* and *Saimiri*: I. Grip type and hand preference. *American Journal of Primatology* 15(3): 235–245.

Crofoot MC, Gilby IC, Wikelski MC, Kays RW (2008). Interaction location outweighs the competitive advantage of numerical superiority in Cebus capucinus intergroup contests. *Proceedings of the National Academy of Sciences of the United State of America* 105: 577–581.

Crofoot, M.C. (2007). Mating and feeding competition in white-faced capuchins (*Cebus capucinus*): the importance of short- and long-term strategies. *Behaviour* 144: 1473–1495.

Cunningham E, Janson C (2007). Integrating information about location and value of resources by white-faced saki monkeys (*Pithecia pithecia*). *Animal Cognition* 10(3): 293–304.

Cunningham EP (2003). The use of memory in *Pithecia pithecia*'s foraging strategy. The City University of New York. PhD dissertation.

Cunningham EP, Janson CH (2001). Effect of small home range size on use of memory. *American Journal of Physical Anthropology* (Suppl. 32): 56.

Cunningham EP, Janson CH (2003). Effect of fruit scarcity on use of memory. *American Journal of Physical Anthropology* (Suppl. 36): 82.

Cunningham EP, Janson CH (2006). *Pithecia pithecia*'s behavioral response to decreasing fruit abundance. *American Journal of Primatology* 68: 491–497.

Darwin Ch. (1873 [1872]). *The expression of the emotions in man and animals.* New York: D. Appleton.

De La Ossa J, Galván Guevara S, Fajardo Patiño A.(2013). Densidad, composición de grupo y distribución vertical de primates simpátricos en un bosque de galería fragmentado, Colosó, Sucre – Colombia. *Revista U.D.C.A Actualidad & Divulgación Científica* 16: 185–192.

De Lillo C, Aversano M, Tuci E, Visalberghi E (1998). Spatial constraints and regulatory functions in monkeys' (*Cebus apella*) search. *Journal of Comparative Psychology* 112: 353–362.

De Lillo C, Visalberghi E, Aversano M (1997. The organization of exhaustive searches in a "patchy" space by capuchin monkeys (*Cebus apella*). *Journal of Comparative Psychology* 111: 82–90.

De Lillo C., Spinozzi G., Truppa V. (2007). Pattern recognition in tufted capuchin monkeys (*Cebus apella*): the role of the spatial organization of stimulus parts. *Behavioural Brain Research* 181: 96–109.

De Luna, AG, Link A (2018). Distribution, population density and conservation of the critically endangered brown spider monkey (*Ateles hybridus*) and other primates of the inter-Andean forests of Colombia. *Biodiversity and Conservation* 27: 3469–3511.

de Steven D, Windson DM, Putz FE, de Leon B (1978. Vegetative and reproductive phenologies of a palm assemblage in Panama. *Biotropica* 19: 342–356.

Defler TR (1976). *An analysis of grooming in two species of macaque (*Macaca nemestrina *and* Macaca radiata*).* Denver: University of Colorado, PhD Dissertation.

Defler TR (1979). On the ecology and behavior of *Cebus albifrons* in Eastern Colombia: I. Ecology. II. Behavior. *Primates* 20. 475–490, 491–502.

Defler TR (1982). A comparison of intergroup behavior in *Cebus albifrons* and *C. apella*. *Primates* 23: 385–392.

Defler TR (2004). *Primates of Colombia*. Bogotá: Conservation International.

Di Bitetti MS (2009). The distribution of grooming among female primates: Testing hypotheses with the Shannon-Wiener diversity index. *Behaviour* 137: 1517–1540.

Di Bitetti MS, Janson CH (2001). Social foraging and the finder's share in capuchin monkeys, *Cebus apella. Animal Behaviour* 62: 47–56.

Di Fiore A, Suárez S (2004). Route-based travel and shared routes in sympatric spider and woolly monkeys. *Folia Primatologica* 75(suppl. 1): 97–98.

Di Fiore A, Suarez SA (2007). Route-based travel and shared routes in sympatric spider and woolly monkeys: cognitive and evolutionary implications. *Animal Cognition* 10: 317–329.

Dolins FL, Mitchell RW. (2010). *Spatial Cognition, Spatial Perception Mapping the Self and Space*. Cambridge, Cambridge University Press.

Dominy N, Lucas P, Osorio D, Yamashita N (2001). The sensory ecology of primate food perception. *Evolutionary Anthropology* 10: 171–186.

Dunbar RIM (1992). A model of the gelada socio-ecological system. *Primates* 33: 69–83.

Dunbar RIM (1994). A general theory of primate social systems: The Osman Hill lecture. *Folia Primatologica* 64: 95–96.

Dunbar RIM (1998. The social brain hypothesis. *Evolutionary Anthropology* 6(5): 178–190.

Duque NJ, Gómez-Posada C. (2009). Sympatric Alouatta seniculus and Cebus capucinus in an Andean Forest Fragment in Colombia: A Survey of Population Density. *Neotropical Primates* 16: 51–56.

Dyer FC (1996). Spatial memory and navigation by *honeybees* on the scale of the foraging range. *Journal of Experimental Biology* 199: 147–154.

Dyer FC (1998). Cognitive ecology of navigation. *In:* R Dukas (ed.), *Cognitive Ecology. The evolutionary ecology of information processing and decision making*. Chicago: Univ Chicago, pp. 201–260.

Dyer FC (2000). Group movements and individual cognition: lessons from social insects. *In:* P. A. Garber and S Boinski (eds.), *On the move: How and why animals travel in groups*. Chicago, University of Chicago Press, pp. 127–164.

Eadie, E.C. (2012). *Feeding ecology and life history strategies of white-faced capuchin monkeys (*Cebus capucinus*)*. Ph.D. Dissertation University of New Mexico, USA.

EBLS (Estación Biológica La Suerte) (2009). La Suerte-Maps. http://lasuerte.org/mapsidx.html (Consulted on January, 2009).

Encarnación F, Castro N. (1990). Informe preliminar sobre censo de primates no humanos en el Sur Oriente Peruano: Iberia e Iñapari (Departamento de Madre de Dios), Mayo 15 – Junio 4, 1978. *In:* Castro-Rodríguez, N. E. (eds.), *La primatología en el Perú, Vol. I*. Universidad Mayor de San Marcos, pp. 122–134.

Erhart EM, Overdolff DJ (2008). Spatial memory during foraging in prosimians primates: *Propithecus edwardsi* and *Eulemur fulvus rufus. Folia Primatologica* 79: 185–196.

Erhart EM, Overdorff DJ (1999). Female coordination of group travel in wild *Propithecus* and *Eulemur. International Journal of Primatology* 20: 927–940.

Etienne AS, Berlie J, Georgakopoulos J, Maurer R (1998). Role of dead reckoning in navigation. *In:* Healy S (ed.): *Spatial prepresentation in animals.* Oxford: Oxford University
Press, pp. 54–68.

Etienne AS, Maurer R, Saucy F (1988). Limitations in the assessment of path dependent
information. *Behaviour* 106: 81–111.

Farrall L. (1979). *Unwritten Knowledge: Case Study of the Navigators of Micronesia.* Deakin
University School of Humanities. Open Campus Program, Open Campus Program.

Fedigan LM (1990). Vertebrate predation in *Cebus capucinus*: Meat eating in a Neotropical
primate. *Folia Primatologica* 54: 196–205.

Fedigan LM, Jack KM (2004). The demographic and reproductive context of male replace-
ments in *Cebus capucinus. Behaviour* 141: 755–775.

Fedigan LM, Rose LM (1995. Interbirth interval variation in three sympatric species of
neotropical monkey. *American Journal of Primatology* 37: 9–24.

Fedigan LM, Rose LM, Avila RM (1996). See how they grow: Tracking capuchin monkey
(*Cebus capucinus*) populations in a regenerating Costa Rican dry forest. *In:* Norconk
MA, Rosenberger AL, Garber PA (eds.), *Adaptive Radiations of Neotropical Primates.*
New York: Plenum Press, pp. 289–307, 543–544.

Fernandes MEB (1991). Tool use and predation of oysters (*Crassostrea rhizophorae*)
by the tufted capuchin, *Cebus apella apella*, in brackish water mangrove swamp.
Primates 32: 529–531.

Fernández VA (2008). Patrones de desplazamiento de aulladores negros y dorados
(*Alouatta caraya*). Buenos Aires: Universidad de Buenos Aires, Bachelor thesis.

Ferrari SF, Lopes MA (1996). Primate populations in eastern Amazonia.. In: Norconk,
M.A., Rosenberger, A.L., Garber, P.A. (eds) *Adaptive Radiations of Neotropical
Primates.* Plenum Press, New York, pp. 53–67.

Ford SM, Davis LC (1992. Systematics and body size: Implications for feeding adaptations
in New World monkeys. *American Journal of Physical Anthropology* 88: 415–468.

Fragaszy D (1986). Time budgets and foraging behavior in wedge-capped capuchins
(*Cebus olivaceus*): Age and sex differences. *In:* Taub DM, King FA (eds.), *Current
Perspectives in Primate Social Dynamics.* New York: Van Nostrand Reinhold Co,
pp. 159–174.

Fragaszy D, Cummins-Sebree S (2005). Relational spatial reasoning by a nonhuman: The
example of capuchin monkeys. *Behavioral Cognitive Neuroscience Reviews* 4: 282–306.

Fragaszy D, Fedigan L, Visalberghi E (2004). *The Complete Capuchin: The Biology of the
Genus* Cebus. New York: Cambridge Univ. Press.

Fragaszy D, Izar P, Visalberghi E, Ottoni EB, Gomes de Oliveira M. (2004).Wild capuchin
monkeys (*Cebus libidinosus*) use anvils and stone pounding tools. *American Journal of
Primatology* 64: 359–366.

Fragaszy D, Johnson-Pynn J, Hirsh E, Brakke K (2003). Strategic navigation of two-dimensional alley mazes: Comparing capuchin monkeys and chimpanzees. *Animal
Cognition* 6: 149–160.

Fragaszy, D.M. (1990). Sex and age differences in the organization of behavior in wedge-
capped capuchins, *Cebus olivaceus. Behavioral Ecology* 1: 81–94.

Freese CH (1975). A census of non-human primates in Peru. *In: Primate censusing studies in Peru and Colombia: Report to the National Academy of Sciences on the activities of project Amro-0719*. Washington, DC: Pan American Health Organization, pp. 17–41.

Freese CH (1977). Food habits of white-faced capuchins Cebus capucinus L. (Primates: Cebidae) in Santa Rosa National Park, Costa Rica. *Brenesia* 10: 43–56.

Freese CH, Oppenheimer JR (1981). The capuchin monkeys, genus *Cebus*. *In:* Coimbra-Filho AF, Mittermeier RA (eds.). *Ecology and Behavior of Neotropical Primates*, Rio de Janeiro: Academia Brasileira de Ciencias, Vol. 1: 331–390.

Galdikas BF, Vasey P (1992). Why are orangutnas so smart?: Ecological and social hypotheses. *In:* Burton FD (ed.), *Social processes and mental abilities in non-human primates. Evidences from longidutinal field studies*. New York. The Edwin Mellen Press. pp. 183–224.

Gallistel CR (1989). Animal cognition: The representation of space, time, and number. *Annual Review in Psychology* 40: 155–89.

Gallistel CR (1990). *The organization of learning*. Cambridge, MA: MIT Press.

Gallister CR, Cramer AE. (1996). Computations on metric maps in mammals: Getting oriented and choosing a multi-destination route. *Journal of Experimental Biology* 199: 211–217.

Ganzhorn JU, Klaus S, Ortmann S, Schmid J (2003). Adaptations to seasonality: Some primate and nonprimate examples. *In:* Kappeler PM, Pereira ME (eds.), *Primate Life Histories and Socioecology*. Chicago: Univ Chicago Press, pp. 132.

Garber PA (1987). Foraging strategies among living primates. *Annual Review of Anthropology* 16: 339–364.

Garber PA (1989). Role of spatial memory in primate foraging patterns: *Saguinus mystax* and *Saguinus fuscicollis*. *American Journal of Primatology* 19: 203–216.

Garber PA (1993). Modeling Monkeys: A comparison of computer-generated and naturally occurring foraging patterns in two species of Neotropical primates. *International Journal of Primatology* 14: 827–852.

Garber PA (1993). Seasonal patterns of diet and ranging in two species of tamarin monkeys: stability versus variability. *International Journal of Primatology* 14: 145–166.

Garber PA (2000). Evidence for the use of spatial, temporal, and social information by primate foragers. *In:* Garber PA, Boinski S (eds.), *On the move: How and why animals travel in groups*. Chicago: University Chicago Press, pp. 261–298.

Garber PA (2004). New perspectives in primate cognitive ecology. *American Journal of Primatology* 62: 133–137.

Garber PA ; Bicca-Marques JC ; O Azevedo-Lopes MAO (2008). Primate cognition: integrating social and ecological information in decision-making. *In:* Estrada A, Garber PA Pavelka MSM, Luecke L (eds.), *New Perspectives in the Study of Mesoamerican Primates: Distribution, Ecology, Behavior, and Conservation*. New York: Springer, pp. 365–385.

Garber PA, Bicca-Marques JC (2007). Field experiments: a critical approach to the study of primate cognition. *A Primatologia No Brasil* 10: 547–563.

Garber PA, Bicca-Marques JC, Azevedo-Lopes MAO. (2009). Primate cognition: integrating social and ecological information in decision-making. *In:* Garber PA, Estrada A, Bicca-Marques JC, Heymann EW, Strier KB (eds.), *South American Primates: Comparative Perspectives in the Study Of Behavior, Ecology, and Conservation*, New York: Springer, pp. 365–385.

Garber PA, Brown E (2004). Wild capuchins (*Cebus capucinus*) fail to use tools in an experimental field study. *American Journal of Primatology* 62: 165–170.

Garber PA, Brown E (2006). Use of landmark cues to locate feeding sites in wild capuchin monkeys (*Cebus capucinus*): an experimental field study. *In:* Estrada A, Garber PA Pavelka MSM, Luecke L (eds.), *New Perspectives in the Study of Mesoamerican Primates: Distribution, Ecology, Behavior, and Conservation.* New York: Springer, pp. 311–332.

Garber PA, Brown E. (2006). Use of landmark cues to locate feeding sites in wild Capuchin monkeys (*Cebus capucinus*): An experimental field study. *In:* Estrada A, Garber PA, Pavelka M, Luecke L (eds.), *New perspectives in the study of Mesoamerican primates: Distribution, ecology, behavior and conservation.* Developments in Primatology: Progress and Prospects, Series editor Tuttle RA. New York: Kluwer, pp. 311–332.

Garber PA, Dolins FL (1996). Testing learning paradigms in the field: Evidence for use of spatial and perceptual information and rule-based foraging in wild moustached tamarins. *In:* Norconk M, Rosenberger AL, Garber PA (eds.), *Adaptive radiation of Neotropical primates.* New York: Plenum Press, pp. 201–216.

Garber PA, Dolins FL. (2014). Primate spatial strategies and cognition: introduction to this special issue. *American Journal of Primatlogy* 76: 393–398.

Garber PA, Hannon B (1993). Modeling monkeys: A comparison of computer generated and naturally occurring foraging patterns in 2 species of Neotropical primates. *International Journal of Primatology* 14: 827–852.

Garber PA, Jelinek PE (2006). A preliminary study of travel routes and spatial mapping in mantled howler monkeys (*Alouatta palliata*). *In:* Estrada A, Garber PA, Pavelka M, Luecke L (eds.), *New perspectives in the study of Mesoamerican primates: Distribution, ecology, behavior and conservation.* Developments in Primatology: Progress and Prospects, Series editor Tuttle RA. New York: Kluwer, pp. 287–310.

Garber PA, Lavalle A (1999). Experimental approaches to the study of primate cognition in natural and near-to-wild field settings. *In:* Garber PA, Leigh S (eds.), *Readings in the Biological Bases of Human Behavior.* Boston: Pearson Custom Publishing, pp. 71–98.

Garber PA, Paciulli LM (1997). Experimental field study of spatial memory and learning in wild capuchin monkeys (*Cebus capucinus*). *Folia Primatologica* 68: 236–254.

Garber PA, Pruetz JD (1995). Positional Behavior in Moustached Tamarin Monkeys: Effects of habitat on locomotor variability and stability. *Journal of Human Evolution* 28: 411–426.

Garber PA, Rehg JA (1999). The ecological role of the prehensile tail in white-faced capuchins (*Cebus capucinus*). *American Journal of Physical Anthropology* 110: 325–339.

Garcés-Restrepo MF, Carvajal-Nieto P, Astorquiza JM, Sánchez-Vélez E, Giraldo A (2014). Density and structure of populations of *Cebus capucinus curtus* (Primates: Cebidae) and *Bradypus variegatus gorgon* (Pilosa: Bradypodidae) in Gorgona Island, Colombia. *Revista de Biología Tropical* 62: 447–456.

Gibson BM, Kamil AC (2001). Test for cognitive mapping in Clark's nutcrackers (*Nucifraga columbiana*). *Journal of Comparative Psychology* 115: 403–407.

Gibson KR (1986). Cognition, brain size and the extraction of embedded food re s. *In:* Else JG, Lee PC (eds.), *Primate Ontogeny, Cognition and Social Behaviour.* New York: Cambridge University Press. pp. 93–103.

Gilissen E (1999). Relative brain size is not independent of body size in primates: Comparison between humans and capuchin monkeys. *American Journal of Physical Anthropology* (Suppl. 28): 134.

Gilissen E (2003). Evaluation du cerveau des primates: implications de la taille du corps (Scaling of the primate brain: implications of body size). *Anthropologie et Préhistoire* 114: 81–84.

Green KM (1978). Primate censusing in northern Colombia: A comparison of two techniques. *Primates* 19, 537–550 .

Groves CP (2001). *Primate Taxonomy*. Washington, DC: Smithsonian Institution Press: 350 p.

Gursky S (2009). Effect of seasonality on the behavior of an insectivorous primate, Tarsius spectrum. *International Journal of Primatology* 21: 477–495.

Hakeen *et al.* (1996). Brain and life span in primates *In:* Birren J (ed.), *Handbook of the Psychology of Aging*. San Diego: Academic Press, pp. 78–104.

Harada ML, Ferrari SF (1996). Reclassification of *Cebus kaapori* Queiroz, 1992 based on new specimens from eastern Para, Brazil. *IPS/ASP Congress Abstracts*, Abstract #729.

Harrison-Levine1 AL, Norconk MA, Cunningham EP (2003). Insect predation techniques suggest predator sensitive foraging in a group of white-faced saki (*Pithecia pithecia*) *American Journal of Primatology* 60(suppl. 1): 66.

Herbinger I, Boesch C, Rothe H (2001). Territory characteristics among three neighboring chimpanzee communities in the Tai National Park, Cote d'Ivoire. *International Journal of Primatology* 22(2): 143–167.

Hernández-Camacho J, Cooper RW (1976). The nonhuman primates of Colombia. *In:* Thorington RW Jr, Heltne PG (eds.), *Neotropical Primates: Field Studies and Conservation*. Washington, DC: National Academy of Sciences, pp. 35–69.

Hershkovitz P (1977). *Living New World Monkeys (Platyrrhini) with an Introduction to Primates*. Chicago: The University of Chicago Press.

Hinde RA (2009). Some reflections on primatology at Cambridge and the science studies debate. *In:* Strum SC, Fedigan LM (eds.), *Primate Encounters. Models of Science, Gender, and Society*. Chicago: The University of Chicago Press, pp. 104–115.

Hladik A, Hladik CM (1969). Trophic relationships between vegetation and primates in the forest of Barro Colorado (Panama). *Terre et La Vie* 23: 25–117.

Hladik CM, Hladik A, Bousset J, Valdebouze P, Viroben G, Delort-Laval J (1971). Le régime alimentaire des primates de l'ile de Barro Colorado. Résultats des analyses quantitatives (Panama). *Folia Primatologica* 16: 85–122.

Hsu MJ, Agoramoorthy G. (1995). Conservation status of primates in Trinidad, West Indies. *Oryx* 30: 285–291.

Hurtado CM, Serrano-Villavicencio J, Pacheco V. (2016). Population density and primate conservation in the Noroeste Biosphere Reserve, Tumbes, Peru. *Revista Peruana de Biología* 23: 151- 158.

IGCR-Instituto Geográfico de Costa Rica. (1984). *Hoja Río Sucio 3447-III*. Topographic map scale 1: 50.000. 2nd. Edition.

Jack KM (2007). The Cebines: toward an explanation of variable social structure. *In:* Campbell CJ, Fuentes A, MacKinnon KC, Panger M, Bearder SK (eds.), *Primates in Perspective*. New York: Oxford Univ Press, pp. 107–123.

Jack KM, Campos FA (2012). Distribution, abundance, and spatial ecology of the critically endangered Ecuadorian capuchin (*Cebus albifrons aequatorialis*). *Tropical Conservation Science* 5: 1730.

Jack KM, Fedigan LM (2004). Male dispersal patterns in white-faced capuchins, *Cebus capucinus*. Part 1: Patterns and causes of natal emigration. Part 2: Patterns and causes of secondary dispersal resources. *Animal Behaviour* 67: 761–769, 771–782.

Jack KM, Fedigan LM (2006). Why be alpha male? Dominance and reproductive success in wild white-faced capuchins (*Cebus capucinus*). *In:* Estrada A, Garber PA, Pavelka M, Luecke L (eds.), *New perspectives in the study of Mesoamerican primates: Distribution, ecology, behavior and conservation.* Developments in Primatology: Progress and Prospects, Series editor Tuttle RA. New York: Kluwer, pp 367–386.

Janmaat KRL, Chapman CA, Zuberbuehler K, Byrne RW (2008). The use of fruiting synchronicity as a conceptual tool by foraging mangabeys. *Primate Eye* 96(Sp CD-ROM iss-IPS 2008, Abst. #434).

Janmaat KRL, Zuberbuhler K, Byrne RW (2004). Do Mangabeys remember fruiting stages? A study on visiting frequencies of and travel speed towards rain forest trees with different fruiting states. *Folia Primatologica* 75(Suppl. 1): 97.

Janson C, Verdolin J (2005). Seasonality of primate births in relation to climate. *In:* Brockman DK, van Schaik CP (eds.), *Seasonality in primates: Studies of living and extinct human and non-human primates.* New York: Cambridge Univiversity Press. pp. 307–350.

Janson CH (1985). Aggressive competition and individual food consumption in wild brown capuchin monkeys (*Cebus apella*). *Behavioral Ecology and Sociobiology* 18: 125–138.

Janson CH (1990a). Social correlates of individual spatial choice in foraging groups of brown capuchin monkeys, *Cebus apella*. Animal Behavior 40: 910–921.

Janson CH (1990b). Ecological consequences of individual spatial choice in foraging groups of brown capuchin monkeys *Cebus apella. Animal Behavior* 40: 922–934.

Janson CH (1996). Towards an experimental socioecology of primates: Examples for Argentine brown capuchin monkeys (*Cebus apella nigritus*). *In:* Norconk M, Rosenberger AL, Garber PA (eds.), *Adaptive radiation of Neotropical primates.* New York: Plenum Press, pp. 309–325.

Janson CH (1998). Experimental evidence for spatial memory in foraging wild capuchin monkeys, *Cebus apella. Animal Behavior* 55: 1129–1143.

Janson CH (2007). Experimental evidence for route integration and strategic planning in wild capuchin monkeys. *Animal Cognition* 10(3): 341–356.

Janson CH, Boinski S (1992). Morphological and behavioral adaptations for foraging in generalist primates: The case of the cebines. *American Journal of Physical Anthropology* 88: 483–498.

Janson CH, Brosnan, SF. (2013). Experiments in primatology: from the lab to the field and back again. *In* Sterling E, Bynum, N, Blair M (eds). *Primate Ecology and Conservation, Techniques in Ecology and Conservation,* Oxford: Oxford University Press.

Janson CH, Byrne R (2007). What wild primates know about resources: Opening up the black box. *Animal Cognition* 10: 357–367.

Janson CH, Di Bitetti MS (1997). Experimental analysis of food detection in capuchin monkeys: Effects of distance, travel speed, and resource size. *Behavioral Ecology and Sociobiology* 41: 17–24.

Janson CH, Terborgh J, Emmons LH (1981). Non-flying mammals as pollinating agents in the Amazonian forest. *Biotropica* 13 (Suppl. 2): 1–6.

Jelinek PE, Garber PA, Bezanson MF, DeLuycker A, Mara TO (2003). A preliminary study of travel routes and spatial mapping in mantled howler monkeys (*Alouatta palliata*). *American Journal of Physical Anthropology* (Suppl. 36): 122.

Jenkins D (1944). Territory as a result of despotism and social organization as shown by geese. *Auk* 61: 30–47.

Jerison HJ (1973). *Evolution of the brain and intelligence.* New York: Academic Press: 482 p.

Kamil AC, Cheng K (2001). Way-finding and landmarks: the multiple-bearings hypothesis. *Journal of Experimental Biology* 204(1): 103–113.

Kappeler PM, Pereira ME (eds.) (2003). *Primate Life Histories and Socioecology.* Chicago: University of Chicago Press: xxiii, 395 p.

Kay RF (1981). The nut-crackers- -A new theory of the adaptations of the Ramapithecinae. *American Journal of Physical Anthropology* 55: 141–151.

Kay RF, Plavcan JM, Glander KE, Wright PC (1988). Sexual selection and canine dimorphism in New World monkeys. *American Journal of Physical Anthropology* 77(3): 385–397.

Knott CD (1998). Changes in orangutan caloric intake, energy balance, and ketones in response to fluctuating fruit availability. *International Journal of Primatology* 19(6): 1061–1079.

Knott CD (2005). Energetic responses to food availability in the great apes: implications for hominin evolution. *In:* Brockman DK, van Schaik CP (eds.), *Seasonality in primates: Studies of living and extinct human and non-human primates.* New York: Cambridge University Press. pp. 351–378.

Kölher W (1925). *The mentality of apes.* New York: Harcourt, Brace & Co. Inc, 342 p.

Kowalewski MM, Blomquist GE, Urbani B (2009). The effect of travel cost on group size: A phylogenetic approach. *American Journal of Physical Anthropology*, 138 (Suppl.): 228.

Kühlhorn F (1939). Beobachtungen über das Verhalten von Kapuzineraffen in freier Wildbahn. *Zeitschift für Tierpsychologie* 147–151.

Kummer H (1968). Social organization of hamadryas baboons. A field study. Source *Bibliotheca Primatologica* 6: 1–189.

Kummer H (1995). *In quest of the sacred Baboon. A scientist's journey.* Princeton: Princeton University Press.

Lambert JE (2007) Primate nutritional ecology: feeding biology and diet at ecological and evolutionary scales. *In:* Campbell C, Fuentes A, MacKinnon KC, Panger M, Bearder S (eds.), *Primates in Perspective.* Oxford University Press. pp 482–495.

Lambert JE (2007) Seasonality, fallback strategies, and natural selection: a chimpanzee versus cercopithecoid model for interpreting the evolution of hominin diet. In Ungar P (ed.), *Evolution of Human Diet: The Known, the Unknown, and the Unknowable.* University of Oxford Press.

Leigh SR (2004). Brain growth, life history, and cognition in primate and human evolution. *American Journal of Primatology* 62: 139–164.

Lemos JR, Rodal MJN (2002). Woody component of caatinga vegetation in the Parque Nacional da Serra da Capivara, Piauí State, Brazil. *Acta Bot. Bras.* 16: 23–42.

Leonard WR, Robertson ML (2000). Ecological correlates of home range variation in primates: Implications for homind evolution. *In:* Garber PA, Boinski S (eds.), *On the move: How and why animals travel in groups*. Chicago: Univ Chicago Press, pp. 628–648.

Lewis D (1976). Observations on route finding and spatial orientation among the aboriginal peoples of the western desert region of central Australia. *Oceania* 46: 249–282.

Liben LS (1981). Spatial representation and behavior: Multiple perspectives. In: Liben LS, Patterson AH, Nerwcombe N (eds.), *Spatial representation and behavior across the life span*. New York: Academic press.

Liben LS (1988). Conceptual issues in the development of spatial cognition. In: Stiles-Davis J, Kritchevsky M, Bellugi U (eds.), *Spatial cognition: Brain bases and development*. Hillsdale, NJ: Lawrence Erlbaum Associatesd, pp. 167–194.

Liben LS, Downs RM (1989). Understanding maps as symbols: The development of map concepts in children. In: Reese HW (ed.), *Advances in child development and behavior*. San Diego: Academic Press, 22: 145–201.

Linares OJ (1998). *Mamiferos de Venezuela.* Caracas: Sociedad Conservacionista Audubon de Venezuela, British Petroleum, 691 p.

Lowe AJ; Sturrock GA (1998). Behaviour and diet of *Colobus angolensis palliatus* Peters, 1868, in relation to seasonality in a Tanzanian dry coastal forest. *Folia Primatologica* 69(3): 121–128.

Lührs ML, Dammhann M, Fitchel C, Kappeler PM 2007. Spatial memory in grey lemurs (*Microcebus murinus*). *Book of Abstracts of the 2nd Congress of the European Federation for Primatology,* p. 67.

MacCabe JT (2000) Patterns and processes of group movement in human nomadic populations: A case study of the Turkana of northwestern Kenya. *In:* Garber PA, Boinski S (eds.), *On the move: How and why animals travel in groups.* Chicago: University of Chicago Press, pp. 649–677.

MacKinnon KC (2002). *Social development of wild white-faced capuchin monkeys (*Cebus capucinus*) in Costa Rica: An examination of social interactions between immatures and adult males.* University of California at Berkeley, PhD dissertation.

MacKinnon KC (2006). Food choice by juvenile capuchin monkeys (*Cebus capucinus*) in a tropical dry forest. *In:* Estrada A, Garber PA, Pavelka M, Luecke L (eds.), *New perspectives in the study of Mesoamerican primates: Distribution, ecology, behavior and conservation.* Developments in Primatology: Progress and Prospects, Series editor Tuttle RA. New York: Kluwer, pp pp. 349–365.

Makwana SC (1979). Field ecology and behaviour of the rhesus macaque, Macaca mulatta. II. Food, feeding and drinking in Dehra Dun forests. *Indian Journal of Forestry* 2: 242–253.

Mallott EK, PA Garber, RS Malhi. (2017). Integrating feeding behavior, ecological data, and DNA barcoding to identify developmental differences in invertebrate foraging strategies in white-faced capuchins (*Cebus capucinus*). *American Journal of Physical Anthropology* 162: 241–254.

Mallott, E.K. (2016). *Social, ecological, and developmental influences on fruit and inver-tebrate foraging strategies and gut microbial communities in white-faced capuchins (Cebus capucinus).* Ph.D. dissertation. University of Illinois at Urbana-Champaign, Urbana USA.

Mann AE (1981). Diet and human evolution. *In:* Harding RSO, Teleki G (eds.), *Omnivorous primates: gathering and hunting in human evolution.* New York: Columbia Univ Press, pp. 10–36.

Marshall AJ, Wrangham RW (2007). Evolutionay consequences of fallback foods. *International Journal of Primatology* 28: 1219–1235.

Martin P, Bateson P (2000). *Measuring behaviour. An Introductory Guide.* Cambridge: Cambridge University Press.

Masterson TJ (2003). Canine dimorphism and interspecific canine form in *Cebus. International Journal of Primatology* 24: 159–178.

Mathews LJ (2008). Ranging behavior of white-fronted capuchins (*Cebus albifrons*) in the

Matthews, L.J. (2009). Activity patterns, home range size, and intergroup encounters in *Cebus albifrons* Support existing models of capuchin socioecology. *International Journal of Primatology* 30: 709–728.

Mauer R, Séguinot V (1995). What is modeling for? A critical review of the models of path integration. Journal of Theoretical Biology, 175: 457–475.

McCabe JJ (2000). Patterns and processes of group movement in human nomadic popu-lations: A case study of the Turkana of northwestern Kenya. *In:* Garber PA, Boinski S (eds.), *On the move: How and why animals travel in groups.* Chicago: Univ Chicago Press, pp. 649–677.

McConkey KR, Ario A, Aldy F, Chivers DJ (2003). Influence of forest seasonality on gibbon food choice in the rain forests of Barito Ulu, Central Kalimantan. *International Journal of Primatology* 24(1): 19–32.

Menzel C (1996). Structure-guided foraging in long-tailed macaques. *American Journal of Primatology* 38: 117–132.

Menzel EW Jr (1991). Chimpanzees (*Pan troglodytes*): Problem seeking versus the bird-in-hand, least-effort strategy. *Primates* 32(4): 497–508.

Michalski F, Michalski LJ, Barnett AP (2017). Environmental determinants and use of space by six Neotropical primates in the northern Brazilian Amazon. *Studies on Neo-tropical Fauna and Environment* 52: 187–197.

Miller L (1992). *Socioecology of the wedge-capped capuchin monkey (Cebus olivaceus).* Davis: University of California, PhD. thesis.

Miller, L.E. (1996). The behavioral ecology of wedge-capped capuchin monkeys (*Cebus olivaceus*). In: Norconk, M.A., Rosenberger, A.L., Garber, P.A. (eds) *Adaptive Radiations of Neotropical Primates.* Plenum Press, New York, pp. 271–288.

Milton K (1980). *The foraging strategy of howler monkeys: A study in primate economics.* New York: Columbia University Press.

Milton K (1981). Distribution patterns of tropical plant foods as an evolutionary stimulus to primate mental development. *American Anthropologist* 83: 534–548.

Milton K (1984). Habitat, diet, and activity patterns of free-ranging woolly spider monkeys (*Brachyteles arachnoides* E. Geoffroy 1806). *International Journal of Prima-tology* 5: 491–514.

Milton K (1988). Foraging behavior and the evolution of primate cognition. *In:* Byrne R W, Whiten A (eds.), *Machiavellian intelligence: Social expertise and the evolution of intellect in monkeys, apes, and humans.* Oxford: Clarendon Press, pp. 285–305.

Milton K (2000). Quo vadis? Tactics of food search and group movement in primates and other animals. *In:* Garber PA, Boinski S (eds.), *On the move: How and why animals travel in groups.* Chicago: University Chicago Press, pp. 375–417.

Milton K, Giacalone J, Wright SJ, Stockmayer G (2005). Do frugivore population fluctuations reflect fruit production? Evidence from Panama. *In:* Dew JL, Boubli JP (eds.), *Tropical Fruits and Frugivores: The Search for Strong Interactors.* Dordrecht: Springer, pp. 5–35.

Milton K, Van Soest PJ, Robertson JB (1980). Digestive efficiencies of wild howler monkeys. *Physiol Zool* 53: 402–409.

Mitchell BJ (1990). *Resources, group behavior, and infant development in white-faced capuchin monkeys,* Cebus capucinus. University of California, Berkeley. PhD Dissertation.

Mithen S (2003). *The prehistory of mind. The cognitive origins of arts and science.* New York: Thames, Hudson.

Mittelstaedt H, Mittelstaedt ML (1982). Homing by path integration. *In:* Papi F, Wallraff HG (eds.), *Avian navigation.* New York: Springer. pp. 290–297.

Mittermeier RA, van Roosmalen MGM (1981). Preliminary observations on habitat utilization and diet in eight Surinam monkeys. *Folia Primatologica* 36: 1–39.

Moffat CB (1903. The spring rivalry of birds, some views on the limit to multiplication. *Irish Naturalist* 12: 152–166.

Mori A (2004a). A study of the use of individual feeding trees over multiple years with special reference to mental maps in Japanese macaques. *Reichorui Kenkyu/Primate Research* (Japan) 20(suppl.): 6. (Abstract in Japanese).

Moscow D, Vaughan C (1987). Troop movement and food habits of white-faced monkeys in a tropical-dry forest. *Revista de Biología Tropical* 35(2): 287–297.

Moura ACA, Lee PC (2004). Capuchin stone tool use in Caatinga dry forest. *Science* 306: 1909.

Muller M, Wehner R (1988). Path integration in desert ants *Cataglyphis fortis*. *Proceedings of the National Academy of Sciences* 85: 5287–5290.

Neville M, Castro N, Marmol A, Revilla J (1976). Censusing primate populations in the reserved area of the Pacaya and Samiria Rivers, Department Loreto, Peru. *Primates* 17: 151–181.

Newcombe NS, Huttenlocher J (2009). *Making space. The development of social representation and reasoning.* Cambridge, Massachusetts: MIT Press.

Nishikawa M (2008). Evidence for spatial knowledge about food re of Japanese macaques (Macaca fuscata) in terms of travelling patterns. *Primate Eye* 96 (Sp CD-ROM ISS-IPS 2008, Abst. #634).

Noser R, Byrne RW 2007. Travel routes and planning of visits to out-of-sight resources in wild chacma baboons, *Papio ursinus. Animal Behaviour* 73: 257–266.

Noser RG, Byrne RW (2004). How do chacma baboons (*Papio hamadryas ursinus*) find the resources they need in the African Savannah? *Folia Primatologica* 75(Suppl. 1): 95–96. (Abstract).

Oatley K. (1974). Mental maps for navigation. *New Scientist*, Dec. 19: 863–866.

Oliveira, S.G. de, Lynch-Alfaro, J.W.,Veiga, L.M. (2014), Activity budget, diet, and habitat use in the critically endangered Ka'apor capuchin monkey (*Cebus kaapori*) in Pará State, Brazil: A preliminary comparison to other capuchin monkeys. *American Journal of Primatology* 76: 919–931.

Oppenheimer JR (1968). *Behavior and ecology of the white-faced monkey, Cebus capucinus, on Barro Colorado Island, C. Z.* University of Illinois at Urbana-Champaign. PhD dissertation.

Oppenheimer JR (1982). *Cebus capucinus*: Home range, population dynamics, and interspecific relationships. *In:* Leigh EG Jr., Rand AS, Windsor DM (eds.), *The ecology of a tropical forest: Seasonal rhythms and long-term changes.* Washington, DC: Smithsonian Institution Press, pp. 253–272.

Ottoni EB, Mannu M (2001). Semifree-ranging tufted capuchins (*Cebus apella*) spontaneously use tools to crack open nuts. *International Journal of Primatology* 22: 347–358.

Overdorff D, Erhart E (2001). Social and ecological influences on female dominance in day-active prosimian primates. *American Journal of Physical Anthropology* S32: 116.

Palacios, E, Peres CA (2005). Primate population densities in three nutrient-poor Amazonian terra firme forests of south-eastern Colombia. *Folia Primatologica* 76: 135–145.

Panger M (1999. Capuchin object manipulation. *In:* Dolhinow P, Fuentes A (eds.), *The Nonhuman Primates.* Mountain View: Mayfield Publ., pp. 115–120.

Panger MA (1998). Object-use in free-ranging white-faced capuchins (*Cebus capucinus*) in Costa Rica. *American Journal of Physical Anthropology* 106: 311–321.

Panger MA, Perry S, Rose L, Gros-Louis J, Vogel E, Mackinnon KC, Baker M (2002). Cross-site differences in foraging behavior of white-faced capuchins (*Cebus capucinus*). *American Journal of Physical Anthropology* 119: 52–66.

Panger MA, Perry S, Rose L, Gros-Louis J, Vogel E, Mackinnon KC, Baker M. (2002). Cross-site differences in foraging behavior of white-faced capuchins (*Cebus capucinus*). *American Journal of Physical Anthropology* 119: 52–66.

Passamani M, Rylands AB (2000). Home range of a Geoffroy's marmoset group, *Callithrix geoffroyi* (Primates, Callitrichidae) in south-eastern Brazil. *Revista Brasileira de Biologia* 60: 275–281.

Penny ND, Arias JR (1982). *Insects of an* Amazon Forest. N.Y.: Columbia Univ. Press 269 p.

Pereira TS (2004). *Comportamento alimentar, padrão de atividades e uso de espaço por um grupo de* Alouatta caraya *(Primates, Atelidae) en um fragmento de mata no município de Barrinha, SP.* Universidade de São Paulo, Departamento de Biologia. Bachelor thesis.

Pereira TS, Marne OG, Minei CC, Fermoseli AFO, Tognon FR, Cabral A, Hirano ZMB, Santos WF (2005). Uso de área e utilização de rotas de locomoção por um grupo de bugios pretos (*Alouatta caraya* Humboldt 1812) en un fragmento de mata no municipio Barrinha/SP. *XI Congresso Brasileiro de Primatologia, Porto Alegre, Libro de Resumos,* p. 149. (Abstract).

Peres CA (1993). Structure and spatial organization of an Amazonian terra firme forest primate community. *Journal of Tropical Ecology* 9, 259–276.

Peres CA (1994). Seed predators in thirty Amazonian primate communities. *XV Congress of the International Primatological Society* p. 365 (Abstract).

Peres CA (2000). Identifying keystone plant resources in tropical forests: the case of gums from *Parkia* pods. *Journal of Tropical Ecology* 16: 287–317.

Perry S (1996). Social relationships in wild white-faced capuchin monkeys, Cebus capucinus. *Dissertation Abstracts International* A56: 3194.

Perry S (2006). What cultural primatology can tell anthropologists about the evolution of culture. *Annual Review of Anthropology* 35: 171–190.

Perry S, Baker M, Fedigan L, Gros-Louis J, Jack K, MacKinnon K, Manson J, Panger M, Pyle K, Rose L (2003). Social conventions in wild white-faced capuchin monkeys-Evidence for traditions in a neotropical primate. *Current Anthropology* 44(2): 241–268.

Perry S, Barrett HC, Manson JH (2004). White-faced capuchin monkeys show triadic awareness in their choice of allies. *Animal Behaviour* 67: 165–170.

Perry S, Manson JH (2003). Traditions in monkeys. *Evolutionary Anthropology* 12: 71–81.

Perry S, Ordóñez-Jiménez JC (2007). The effects of food size, rarity, and processing complexity on white-faced capuchins' visual attention to foraging conspecifics. In. Hohmann G, Robbins MM, Boesch C (eds.), *Feeding Ecology in Apes and Other Primates: Ecological, Physical and Behavioral Aspects*. New York: Cambridge Univ Press, pp. 203–234.

Phillips KA, Abercrombie CL. (2003). Distribution and conservation status of the primates of Trinidad. *Primate Conservation* 19: 19–22.

Phillips O, Miller JS (2002). *Global patterns of plant diversity: Alwyn H. Gentry's forest transect data set*. St. Louis, Mo.: Missouri Botanical Press, Monographs in systematic botany from the Missouri Botanical Garden, v. 89, xvi + 319 p.

Poole RW (1974). *An introduction to quantitative ecology*. New York: McGraw-Hill.

Pope TR (2009). The evolution of male philopatry in neotropical monkeys. *In:* Kappeler PM (eds.), *Primate Males: Causes and Consequences of Variation in Group Composition*. Cambridge: Cambridge Univ Press, pp. 219–235.

Porter LM, Garber PA, Nacimento E (2009). Exudates as a fallback food for *Callimico goeldii*. *American Journal of Primatology* 71: 120–129.

Potì P (1996). Spatial constructions by capuchin monkeys. *Annali dell'Istituto Superiore di Sanità* 32: 361–367.

Potì P (2000). Aspects of spatial cognition in capuchins (*Cebus apella*): Frames of references and scale of space. *Animal Cognition* 3: 67–77.

Potì P, Bartolommei P, Saporti M (2004). No configurational use of landmarks by capuchins. *Folia Primatologica* 75(Suppl. 1): 319.

Potì P, Bartolommei P, Saporti M (2005). Landmark use by *Cebus apella*. *International Journal of Primatology* 26: 921–948.

Potts R (1998). Environmental hypotheses of hominid evolution. *Yearbook of Physical Anthropology* 41: 93–136.

Potts R (2004). Paleoenvironmental basis of cognitive evolution in great apes. *American Journal of Primatology* 62: 209–228.

Poucet B (1993). Spatial cognitive maps in animals: New hypotheses on their structure and neural mechanisms. *Psychological Review* 100: 163–82.

Ramos-Fernández G, Mateos JL, Miramontes O, Cocho G, Larralde H, Ayala-Orozco B (2004). Lévy walk patterns in the foraging movements of spider monkeys (*Ateles geoffroyi*). *Behavioral Ecology and Sociobiology* 55: 223–230.

Renfrew C (1998). All the King's Horses. *In:* Mithen S (ed.), *Creativity in Human Evolution and Prehistory*. New York and London: Routledge, pp. 260–284.

Roberts WA, Mitchell S, Phelps MT (1993). Foraging in laboratory trees: Spatial memory in squirrel monkeys. *In:* Zentall TR (ed.), *Animal Cognition. A tribute to Donald A. Riley*. Hillsdale, NJ: Lawrence Erlbaum Associates, pp. 131–155.

Robinson JG (1986). Seasonal variation in use of time and space by the wedge-capped capuchin monkey, *Cebus olivaceus*: Implications for foraging theory. *Smithsonian Contributions to Zoology* 431: 1–60.

Robinson JG (1988. Demography and group structure in wedge-capped capuchin monkeys, *Cebus olivaceus*. *Behaviour* 104(3–4): 202–232.

Robinson JG, Janson CH (1987). Capuchins, squirrel monkeys, and atelines: Socioecological convergence with Old World primates. *In:* Smuts DL, Cheney RM, Seyfarth RW, Wrangham TT, Struhsaker BB (eds.), *Primate Societies*. Chicago: University of Chicago Press, pp. 69–82.

Rose LM (1997). Vertebrate predation and food-sharing in *Cebus* and *Pan*. *International Journal of Primatology* 18(5): 727–765.

Rose LM (1998). *Behavioral ecology of white-faced capuchins* (Cebus capucinus*) in Costa Rica*. St Louis, Missouri: Washington University, PhD dissertation.

Rose LM, Perry S, Panger MA, Jack K, Manson JH, Gros-Louis J, Mackinnon KC, Vogel E (2003). Interspecific interactions between *Cebus capucinus* and other species: Data from three Costa Rican sites. *International Journal of Primatology* 24: 759–796.

Rowe N (1996). *The Pictorial Guide to the Living Primates*. East Hampton, NY: Pogonias Press, 263 p.

Ryland AB (2001). Two taxonomies of the New World primates – A comparison of Rylands et al. (2000) and Groves (2001). *Neotropical Primates* 9: 121–124.

Ryland AB (2004). Taxonomic issues and the diversity of neotropical primates. *Folia Primatologica* 75(Suppl. 1): 203 (Abstract).

Rylands AB, Groves CP, Mittermeier RA, Cortes-Ortiz L, Hines JJH (2006). Taxonomy and distributions of Mesoamerican primates. *In:* Estrada A, Garber PA, Pavelka M, Luecke L (eds.), *New perspectives in the study of Mesoamerican primates: Distribution, ecology, behavior and conservation*. Developments in Primatology: Progress and Prospects, Series editor Tuttle RA. New York: Kluwer, pp 29–79.

Rylands AB, Schneider H, Langguth A, Mittermeier RA, Groves CP, Rodriguez-Luna E (2000). An assessment of the diversity of New World primates. *Neotropical Primates* 8: 61–93.

Sampaio Everardo VSB, Silva Grécia C (2005). Equações para estimar a biomassa de plantas da caatinga do semi-árido brasileiro. *Acta Botanica Brasilica* 19: 935–943.

Schooler LJ, Serio-Silva JC, Rhine R (2000a). Does human memory reflect the environment of early hominids.? *CogSci2000 Conference, Institute for Research in Cognitive Science at the University of Pennsylvania*. Philadelphia, August 13–15. http: //www.ircs.upenn. edu/cogsci2000/PRCDNGS/SPRCDNGS/abstrcts/schoosrh. pdf (Abstract).

Schooler LJ, Serio-Silva JC, Rhine R (2000b). Does ACT-R's activation equations reflect the environment of early hominids.? *Seventh Annual ACT-R Workshop. Carnegie Mellon University*. August 5–7 2000. http://act-r.psy.cmu.edu/workshops/Workshop-2000/talks/Schooler.pdf (Abstract).

Schooler LJ, Serio-Silva JC, Rhine R. (no date). *Reflections of early hominid environments in human memory: an analysis of primate ranging behavior*. Unpublished manuscript.

Schuck-Paim C, Kacelnik A (2007). Choice processes in multialternative decision making. *Behavioral Ecology* 18: 541–550.

Serio-Silva JC, Rico-Gray V, Hernández-Salazar LT, Espinosa-Gómez R. (2002). The role of *Ficus* (Moraceae) in the diet and nutrition of a troop of Mexican howler monkeys, *Alouatta palliata mexicana*, released on an island in southern Veracruz, Mexico. *Journal of Tropical Ecology* 18: 913–928.

Shaffer CA (2004). Foraging, ranging, and spatial memory in the mantled howler monkey (*Alouatta palliata*). *American Journal of Physical Anthropology* S38: 179 (Abstract).

Shettleworth SJ (1998). *Cognition, Evolution, and Behavior*. New York: Oxford Univ Press: 688 p.

SIDBAAP-Sistema de Información de la Diversidad Biológica y Ambiente de la Amazonia Peruana (2009). *Pijuayo*. http://www.siamazonia.org.pe/.

Sigg H (1986). Ranging patterns in hamadryas baboon: Evidence for a mental map. *In:* Else JG, Lee PC (ed.), *Primate ontogeny, cognition and social behaviour. Selected Proceedings of the Tenth Congress of the International Primatological Society, Kenya 1984*. Cambridge: Cambridge University Press, Volume 3: 87–91.

Sigg H, Stolba A. (1981). Home range and daily march in a Hamadryas Baboon troop. *Folia Primatologica* 36: 40–75.

SINAE – Sistema Nacional de Áreas de Conservación (2001). *Costa Rica. Áreas silvestres protegidas y áreas de conservación*. MINAE (Ministerio del Ambiente y Energía), Proyecto ECOMAPAS-INBIO and Norway Goverment. Scale 1: 650 000.

Soares-Bortolini TS, Bicca-Marques JC (2007). A case of spontaneous tool-making by a captive capuchin monkey. *Neotropical Primates* 14: 74–76.

Soini P (1986). A synecological study of a primate community in the Pacaya-Samiria National Reserve, Peru. *Primate Conservation* 7: 63–71.

Soini P (1993). The ecology of the pygmy marmoset, *Cebuella pygmaea*: Some comparisons with two sympatric tamarins. *In:* Rylands AB (ed.), *Marmosets and Tamarins: Systematics, Behaviour, and Ecology*. Oxford: Oxford University Press, pp. 257–261.

Sokal RR, Rohlf FJ (1995). *Biometry*. San Francisco: Freeman. 3rd ed.

Sollins P, Sancho MF, Mata CR, Sanford RL. (1994). Soils and soil process research. *In:* McDade LA, Bawa KS, Hespenheide H, Hartshorn GS (eds.), *La Selva: ecology and natural history of a neotropical rainforest*. Chicago: University of Chicago Press, pp. 34–53.

Southwick CH (1967). An experimental study of intragroup agonistic behavior in rhesus monkeys (*Macaca mulatta*). *Behaviour* 28: 182–209.

Spetch ML, Cheng K, MacDonald SE (1996). Learning the configuration of a landmark array, I: Touch screen studies with pigeons and humans. *Journal of Comparative Psychology* 110: 55–68.

Spinozzi G (2007). Factors affecting manual laterality in tufted capuchins (*Cebus apella*). *In:* Hopkins WD (ed.), *The Evolution of Hemispheric Specialization in Primates*. San Diego: Academic Press, pp. 205–226.

Spinozzi G, De Lillo C, Truppa V (2003). Global and local processing of hierarchical visual stimuli in tufted capuchin monkeys (*Cebus apella*). *Journal of Comparative Psychology*, 117: 15–23.

Spinozzi G, Truppa V, Lagana T (2004). Grasping behavior in tufted capuchin monkeys (Cebus apella): grip types and manual laterality for picking up a small food item. *American Journal of Physical Anthropology* 125: 30–41.

Stephan H, Baron G, Frahm HD (1988). Comparative size of brains and brain components. *In:* Steklis HD, Erwin J (eds.), *Comparative Primate Biology, Volume 4: Neurosciences*. New York: Alan R Liss Inc, pp. 1–38.

Steriade M, Jones EG, McCormick DA (1997). *Thalamus*. Amsterdam and New York: Elsevier, 1679 pp.

Stone AI (2007). Responses of squirrel monkeys to seasonal changes in food availability in an eastern Amazonian forest. *American Journal of Primatology* 69: 142–157.

Suárez SA (2003). *Spatio-temporal foraging skills of white-bellied spider monkeys (*Ateles belzebuth belzebuth*) in the Yasuní National Park, Ecuador*. State University of New York at Stony Brook. PhD dissertation.

Sussman RW (1987). Species-specific dietary patterns in primates and human dietary adaptations. *In:* Kinzey WG (ed.), *The Evolution of Human Behavior: Primate Models*. Albany: State University of New York Press, pp. 151–179.

Sussman RW (2000). Piltdown man, the father of American field primatology. *In:* Strum SC, Fedigan LM (eds.), *Primate Encounters. Models of Science, Gender, and Society*. Chicago: The University of Chicago Press, pp. 85–103.

Sussman RW, Garber PA, Cheverud JM (2005). Importance of cooperation and affiliation in the evolution of primate sociality. *American Journal of Physical Anthropology* 128(1): 84–97.

Swedell L, Hailemeskel G, Schreier A (2008). Composition and seasonality of diet in wild hamadryas baboons: preliminary findings from Filoha. *Folia Primatologica* 79: 476–490.

Teleki G (1981). C. Raymond Carpenter, 1905–1975. *American Journal of Physical Anthropology* 56: 383–385.

Terborgh J (1983). *Five New World Primates: A Study in Comparative Ecology*. Princeton, New Jersey: Princeton University Press, xiv, 260 p.

Terborgh J. (1986). Keystone plant resources in the tropical forests. *In:* Soule ME (ed.), *Conservation biology: the science of scarcity and diversity*. Sinauer: Sunderland, pp. 330–344.

Thorington RW Jr (1967). Feeding and activity of Cebus and Saimiri in a Columbian forest. *In:*. Starck D. Schneider R, Kuhn H-J (eds.), *Neue Ergebnisse der Primatologie*. Stuttgart: Gustav Fischer Verlag. pp. 180–184

Tinsley-Johnson E, Benítez ME, Fuentes A, McLean CR, Norford AB, Ordoñez JC, Beehner JC, Bergman TJ. (2020). High density of white-faced capuchins (Cebus capucinus) and habitat quality in the Taboga Forest of Costa Rica. *American Journal of Primatology* 82: e23096.

Tomasello M, Call J (1997). *Primate cognition*. New York/Oxford: Oxford University Press.

Tournon J, Alvarado G (1995). *Carte géologique – Mapa geológico de Costa Rica*. Dieppe, France: Imprimerie La Vigie, Scale 1: 500 000.

Urbani B (1999). Spontaneous use of tools by wedge-capped capuchin monkeys (*Cebus olivaceus*). *Folia Primatologica* 70: 172–174.

Urbani (2004). *Spatial mapping by wild capuchin monkeys and its implication for the conservation of a Costa Rican rainforest*. Center of Latin American and Caribbean Studies at the University of Illinois at Urbana-Champaign. Unpublished technical report.

Urbani, B. (2009). *Spatial mapping in wild white-faced capuchin monkeys (*Cebus capucinus*)*. Ph.D. dissertation. University of Illinois at Urbana-Champaign, Urbana USA.

Urbani B Garber P. (2002). A stone in their hands… Are monkeys tool users?. *Anthropologie* (Brno) 40: 183–191.

Urbani B, Kowalewski M. (2021). Filogenia y comportamiento: implicaciones en estudios primatológicos. *Estudios de Antropología Biológica* XIX: 175–209.

Urquiza-Haas T, Serio-Silva JC, Hernandez-Salazar LT (2008). Traditional nutritional analyses of figs overestimates intake of most nutrient fractions: a study of Ficus perforata consumed by howler monkeys (*Alouatta palliata mexicana*). *American Journal of Primatology* 70: 432–438.

Valero A, Byrne RW (2003). Do spider monkeys have mental maps? A new procedure to study animal movements in the wild. *PSGB Spring Meeting 2003*, 10–11 April. http://www.psgb.org/Meetings/Spring2003.html (Abstract).

Valero A, Byrne RW (2004). How do spider monkeys find the resources they need in Mexican dry woodland? *Folia Primatologica* 75: 95 (Abstract).

Valero A, Byrne RW (2007). Spider monkey ranging patterns in Mexican subtropical forest: do travel routes reflect planning? *Animal Cognition* 10: 305–315.

van Schaik CP, Brockman DK (2005). Seasonality in primate ecology, reproduction, and life history: an overview. *In:* Brockman DK, van Schaik CP (eds.), *Seasonality in primates: Studies of living and extinct human and non-human primates.* New York: Cambridge Univiversity Press. pp. 3–20.

Ventura V (2004). *Patrón comportamental, alimentación y estrategias de optimización del forrajeo en el mono aullador negro y dorado (*Alouatta caraya*) en el noreste argentino*. Universidad de la República, Facultad de Ciencias, Instituto de Biología, Sección Etología. Montevideo, Uruguay. Bachelor thesis.

Ventura V (2005). Foraging optimization strategies in the black-and-gold howler monkey (*Alouatta caraya*) in Northeast Argentina. *XI Congresso Brasileiro de Primatologia*, Porto Alegre, *Libro de Resumos* p. 149 (Abstract).

Vilela SL, de Faria DS (2004). Seasonality of the activity pattern of *Callithrix penicillata* (Primates, Callitrichidae) in the cerrado (scrub savanna vegetation). *Brazilian Journal of Biology* 64: 363–370.

Visalberghi E (1990). Tool use in *Cebus. Folia Primatologica* 54: 146–153.

Visalberghi E, Limongelli L (1994. Lack of comprehension of cause-effect relations in tool-using capuchin monkeys (*Cebus apella*). *Journal of Comparative Psychology* 108: 15–22.

Visalberghi E, McGrew WC (1997). *Cebus* meets *Pan. International Journal of Primatology* 18: 677–681.

Vogel, E.R. (2005). Rank differences in energy intake rates in white-faced capuchin monkeys, *Cebus capucinus*: the effects of contest competition. *Behavioral Ecology and Sociobiology* 58, 333–344.

Washburn SL (1953). The strategy of physical anthropology. *In*: A. L. Kroeber (ed.), *Anthropology today.* Chicago, University of Chicago Press: 714–727.

Washburn SL (1973). The promise of physical anthropology. *American Journal of Physical Anthropology* 38: 177–182.

Wehner R, Srinivasan MV (1981). Searching behavior in desert ants, genus *Cataglyphis* (Formicidae, Hymenoptera). *Journal of Comparative Physiology* 142: 315–338.

Wehner R, Wehner S (1990). Insect navigation: Use of maps or Ariadne's thread? *Ethology, Ecology and Evolution* 2: 27–48.

White FJ (1998). The importance of seasonality in primatology. *International Journal of Primatology* 19(6): 925–927.

Wolfheim JH (1983). *Primates of the world. Distribution, abundance, and conservation.* Seattle: University of Washington Press, xxiii, 831 p.

Wright BW (2003). The critical function of the "robust" jaws of tufted capuchins. *American Journal of Physical Anthropology* (Suppl. 36): 228 (Abstract).

Wright BW (2004). Food mechanical properties and niche partitioning in a community of Neotropical primates. *American Journal of Physical Anthropology* (Suppl. 38): 212. (Abstract).

Wynn T. (2010). The evolution of human spatial cognition. *In*: Dolins FL, Mitchell RW (eds.). *Spatial Cognition, Spatial Perception Mapping the Self and Space.* Cambridge, Cambridge University Press.

Yamashita N (1996). Seasonality and site specificity of mechanical dietary patterns in two Malagasy lemur families (Lemuridae and Indriidae). *International Journal of Primatology* 17(3): 355–387.

Yerkes RM (1925). *Almost human.* New York: Century, 278 p.

Youlatos D (1998). Positional behavior of two sympatric guianan capuchin monkeys, the brown capuchin (*Cebus apella*) and the wedge-capped capuchin (*Cebus olivaceus*). *Mammalia* 62: 351–366.

Appendix: The Study Area at La Suerte Biological Field Station (EBLS)

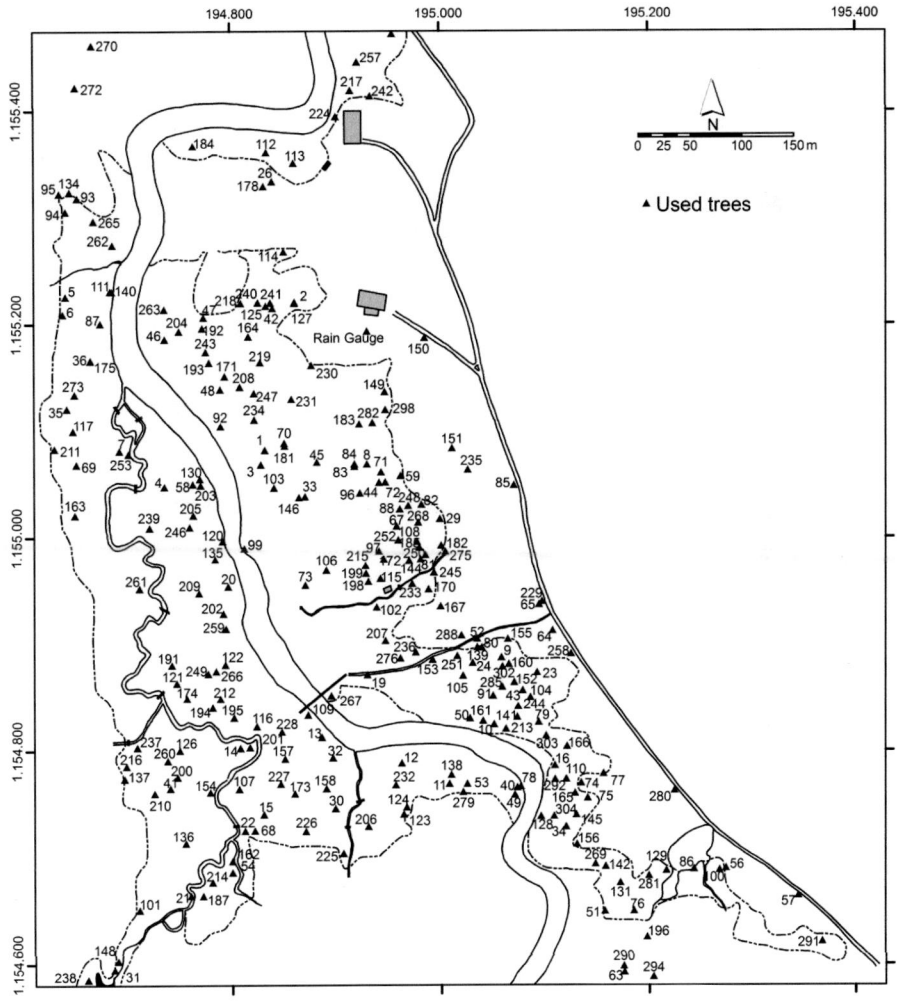

Feeding/Resting trees used by the La Yunai group in the "Small Forest" at the EBLS.

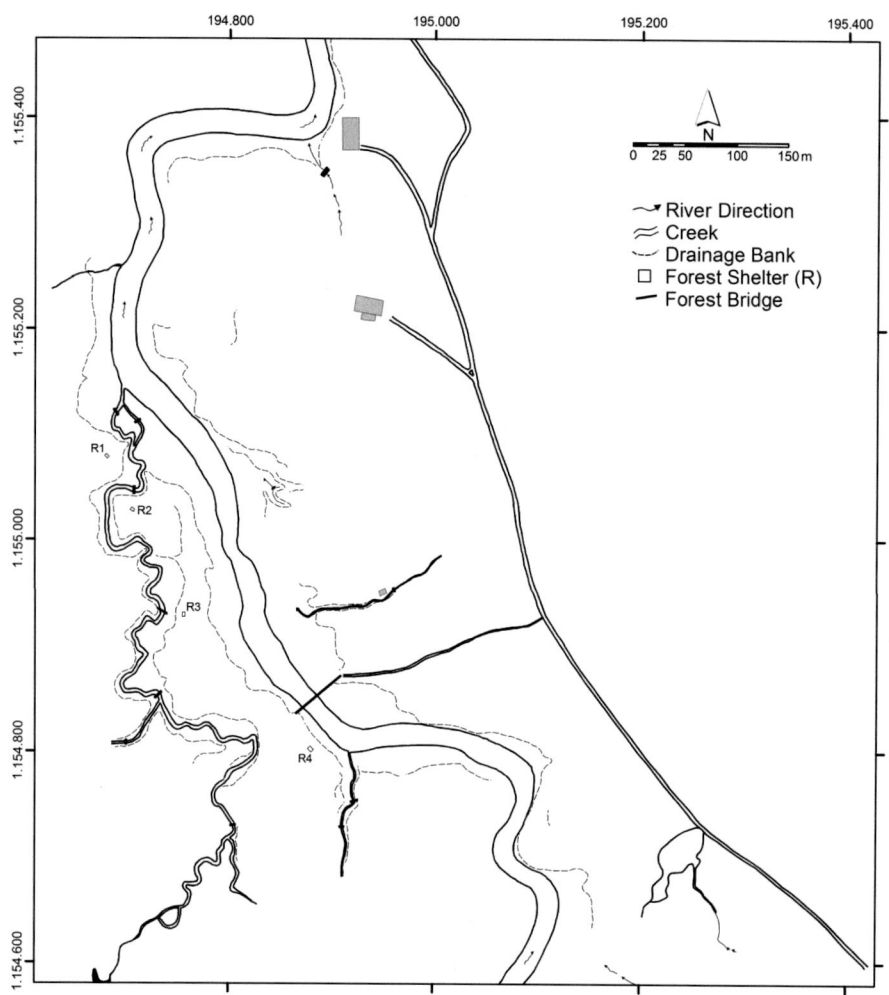

Physical features of the "Small forest" at the EBLS.

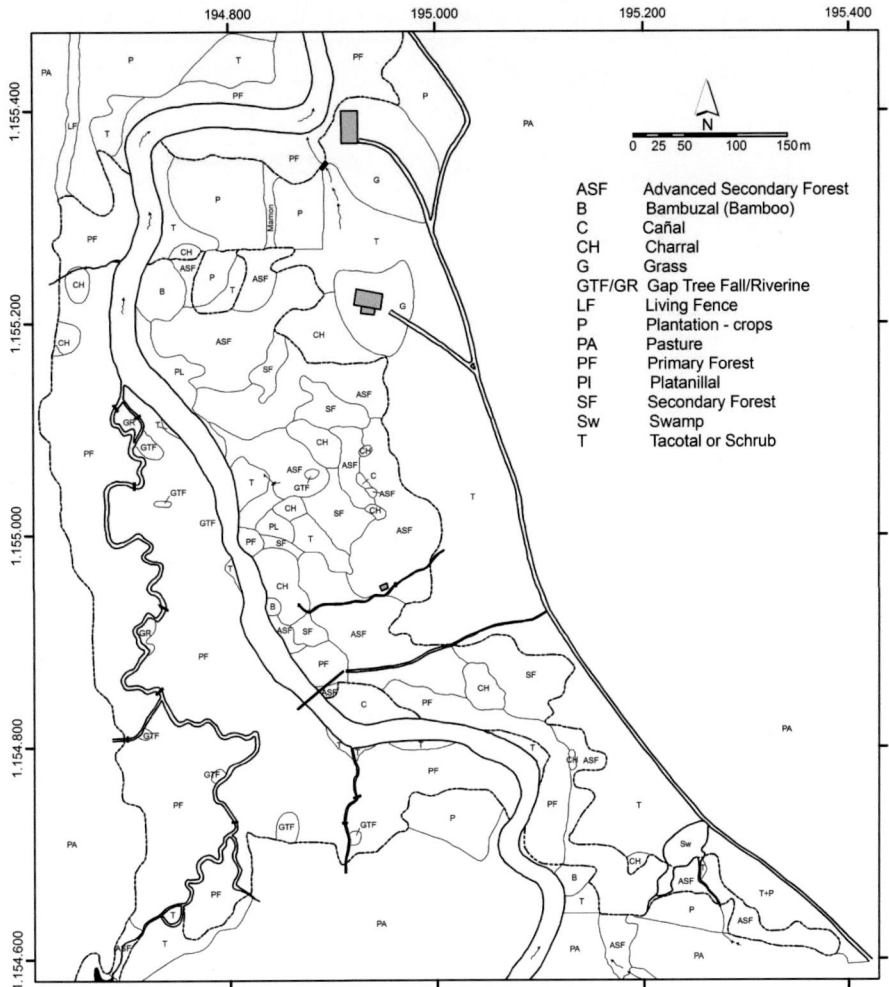

Vegetation types of the "Small Forest" at the EBLS.

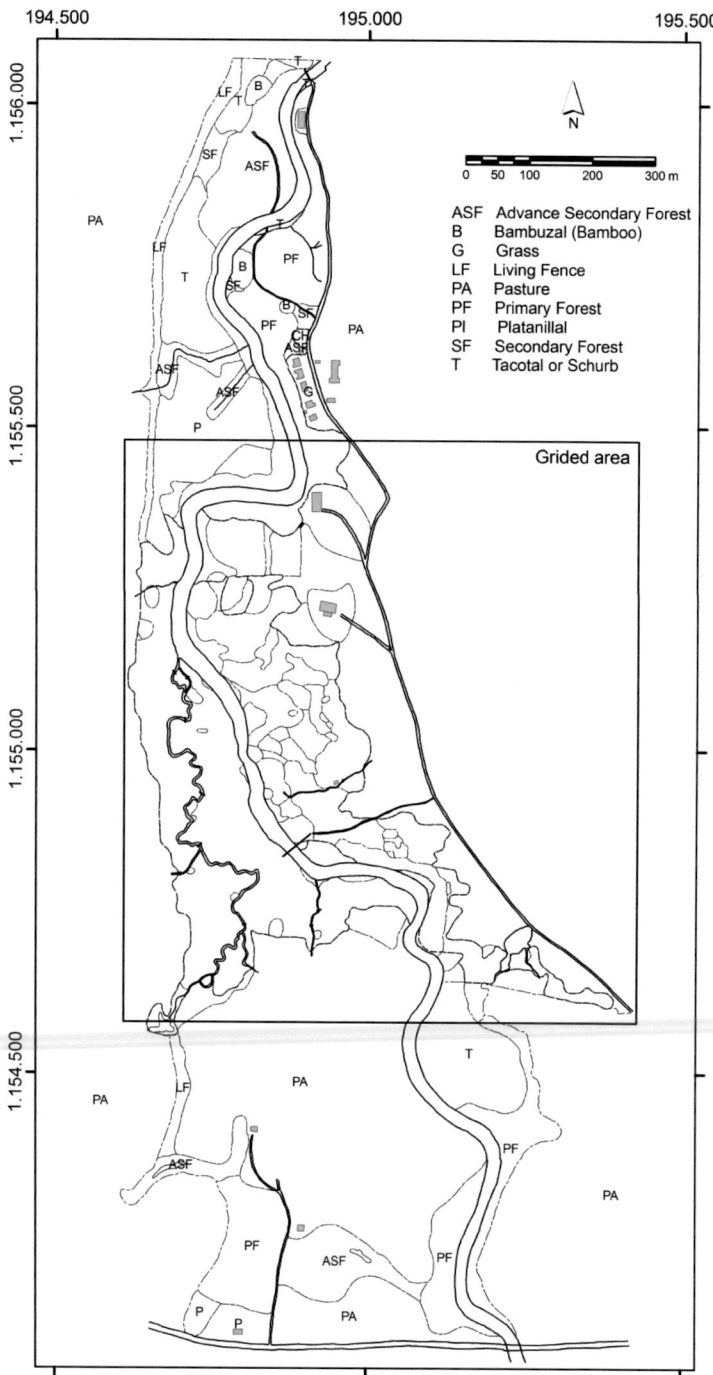

Complete area used by the La Yunai group and vegetation types within and outside the "Small Forest" of the EBLS (the rectangle represents the area covered in the third figure of this appendix).

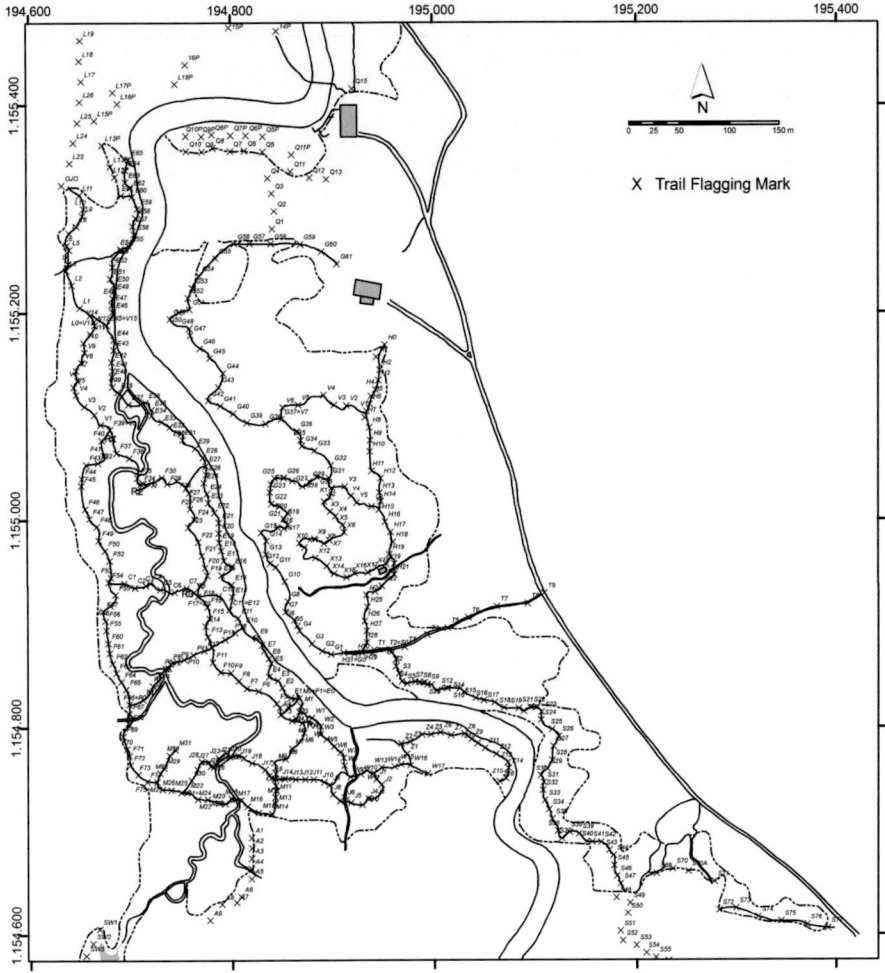

Trail system of the "small forest" at the EBLS (Small Forest, Estación Biológical La Suerte. Provincia de Limón, República de Costa Rica. Survey by Bernardo Urbani, Franco Urbani and Jonathan Mesen-Rubí. January-February 2006).

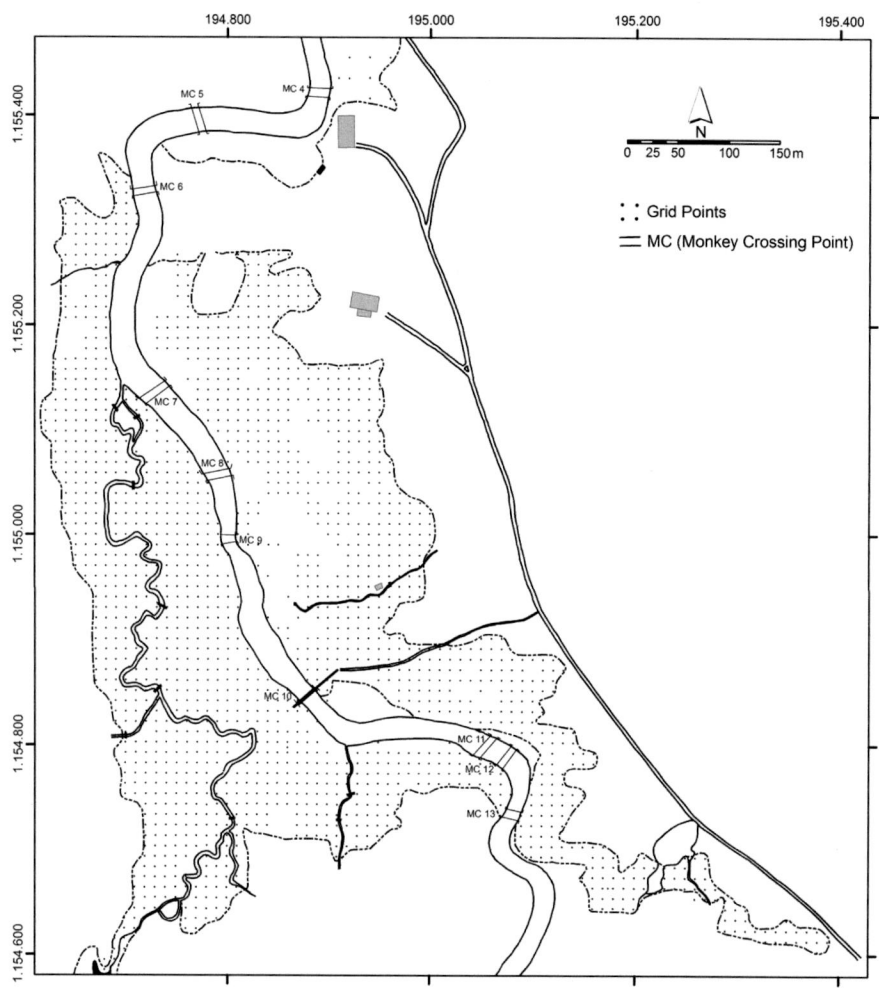

Grid system at the "Small Forest" at the EBLS.

NAVIGATING WITH WHITE-FACED CAPUCHIN MONKEYS

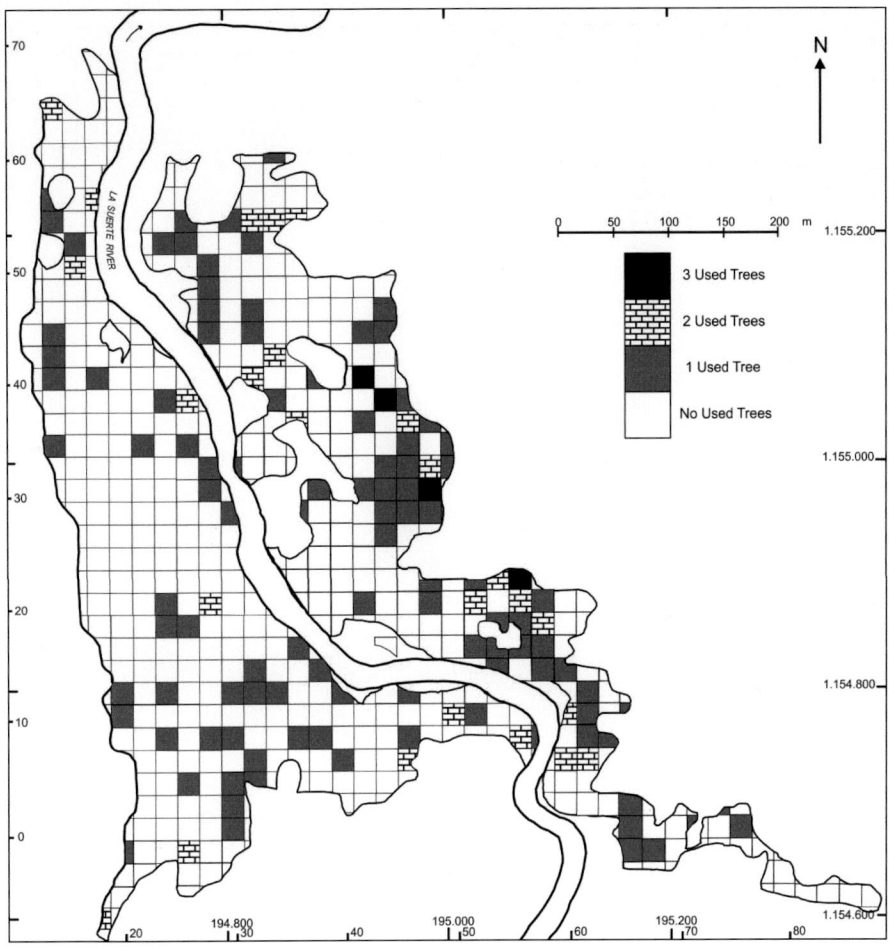

Location of feeding/resting trees within the "Small Forest" at the EBLS (20 m x 20 m boxes).

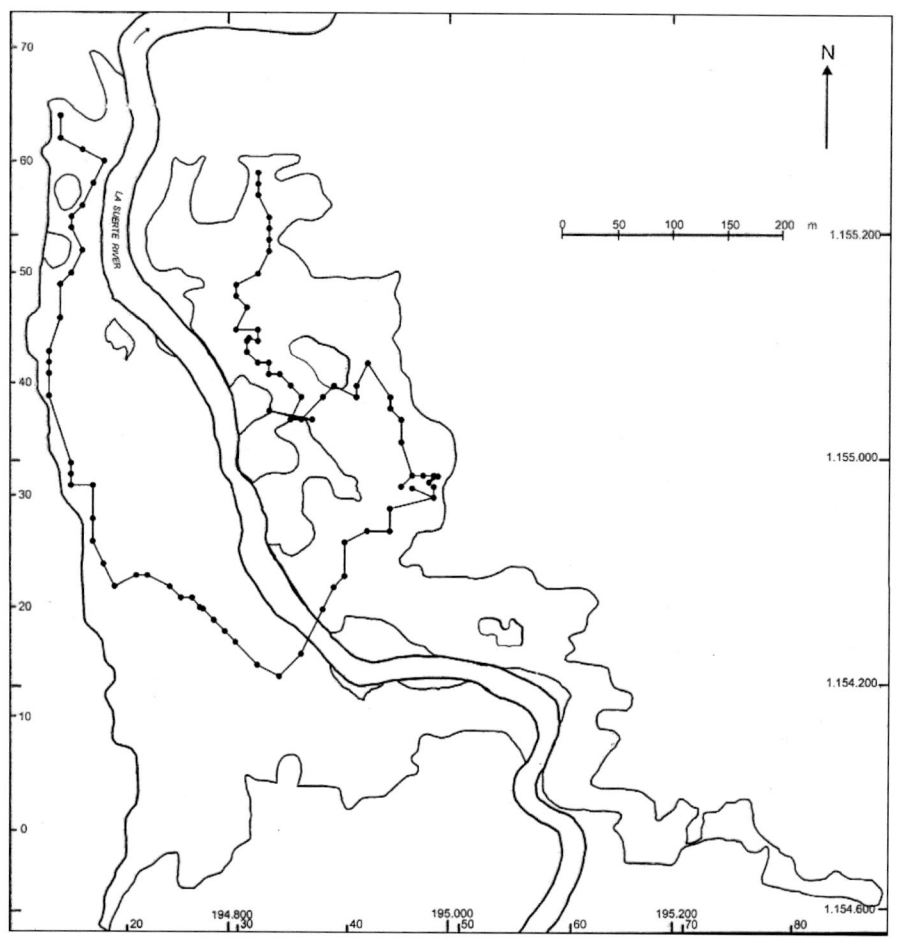

Example of routes of the La Yunai group during the natural field study: October 2006, Day 10.

NAVIGATING WITH WHITE-FACED CAPUCHIN MONKEYS